METHODS IN MOLECULAR BIOLOGY

Series Editor
John M. Walker
School of Life and Medical Sciences
University of Hertfordshire
Hatfield, Hertfordshire, AL10 9AB, UK

For further volumes:
http://www.springer.com/series/7651

B Cell Receptor Signaling

Methods and Protocols

Edited by

Chaohong Liu

Department of Pathogen Biology, Huazhong University of Science and Technology, Huazhong, China

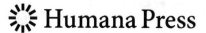

Humana Press

Editor
Chaohong Liu
Department of Pathogen Biology
Huazhong University of Science and Technology
Huazhong, China

ISSN 1064-3745   ISSN 1940-6029   (electronic)
Methods in Molecular Biology
ISBN 978-1-4939-8497-8   ISBN 978-1-4939-7474-0   (eBook)
https://doi.org/10.1007/978-1-4939-7474-0

Printed on acid-free paper

This Humana Press imprint is published by Springer Nature
The registered company is Springer Science+Business Media, LLC
The registered company address is: 233 Spring Street, New York, NY 10013, U.S.A.

Preface

The B cell receptor (BCR) is critical for the function of B cells including cell survival, proliferation, and antibody secretion. It is a complex that is composed of two parts: a membrane-bound immunoglobulin molecule and a signal transduction moiety called Ig-α/Ig-β, bound together by disulfide bridges. When membrane-bound immunoglobulin interacts with an antigen, it induces receptor oligomerization and consequent signal transduction. Later the receptor goes through endocytosis for antigen processing and presentation to helper T cells. Therefore the BCR can transduce the antigen strength from outside into signal inside to maintain the normal function of B cells, and the defect in BCR signaling may cause different immune diseases such as immunodeficiency and autoimmune disease.

During the past few years, the details and underlying mechanisms of BCR signaling are not unveiled because of the limitations in the technique relying only on the biochemical assay. Nowadays with the development of all kinds of high resolution microscopy, flow cytometry, and advanced assays, the event of BCR signaling has been dissected gradually as well as the B cell behavior. This volume of Methods in Molecular Biology focuses on various aspects of B cells. Chapter 1 provides protocols to study the mutant of BCR repertoire to understand the antibody evolution. Chapters 2 and 3 present protocols to learn the interactions between B cells and viruses. Chapters 4 and 5 provide technical examples to learn the mechanical force during BCR activation. Chapters 6 and 7 outline the procedures to study B cells proliferation and signaling by using flow cytometry. Chapter 8 establishes the procedures to study the B cell proliferation and differentiation mediated by Tfh cells. Chapters 9–15 numerate all kinds of microscopy such as confocal microscopy, total internal reflection microscopy, and intravital two-photon microscopy. Chapter 16 offers the photo-activatable antigen presentation system to understand BCR signaling. Chapters 17 and 18 present the methodologies to study the critical cell components related to B cell activation including lipid rafts and actin cytoskeleton.

I have to admit that this volume might not provide complete methods to study BCR signaling and B cell behavior during B cell activation. There are many new methods coming out to study BCR signaling and it requires many volumes to cover this topic, and I have tried to offer a glimpse of the current advanced protocols. I hope the scientific community working in the B cell field can benefit from the protocols published in this volume. I really appreciate all the contributors for their time and experience as well as their great suggestions to form this volume. John Walker, the senior editor, has provided huge support for publishing these protocols, and finally, the staffs from Springer and Jingwen Li also offered great help to edit this volume.

Huazhong, China *Chaohong Liu*

Contents

Contributors

CHRISTOPHER D.C. ALLEN • *Cardiovascular Research Institute, University of California, San Francisco, San Francisco, CA, USA; Sandler Asthma Basic Research Center, University of California, San Francisco, San Francisco, CA, USA; Department of Anatomy, University of California, San Francisco, San Francisco, CA, USA*

LILI AN • *Key Laboratory for Protein and Peptide Pharmaceuticals, Institute of Biophysics, Chinese Academy of Sciences, Beijing, China*

MADISON BOLGER-MUNRO • *Department of Microbiology and Immunology and the Life Sciences Institute, University of British Columbia, Vancouver, BC, Canada*

JOSEPH BRZOSTOWSKI • *Laboratory of Immunogenetics Imaging Facility, National Institute of Allergy and Infectious Diseases, National Institutes of Health, Rockville, MD, USA*

YOLANDA R. CARRASCO • *B Cell Dynamics Laboratory, Department of Immunology and Oncology, Centro Nacional de Biotecnología (CNB)-CSIC, Darwin 3, UAM-Campus Cantoblanco, Madrid, Spain*

AKANKSHA CHATURVEDI • *Department of Biology, Indian Institute of Science Education and Research, Pune, Maharashtra, India*

CHUAN CHEN • *Key Laboratory for Protein and Peptide Pharmaceuticals, Institute of Biophysics, Chinese Academy of Sciences, Beijing, China*

XIN GAO • *China-Australia Centre for Personalised Immunology, Renji Hospital, Shanghai Jiaotong University School of Medicine, Shanghai, China*

DAVID A. GARCIA • *Department of Physics, University of Maryland, College Park, MD, USA*

MICHAEL R. GOLD • *Department of Microbiology and Immunology and the Life Sciences Institute, University of British Columbia, Vancouver, BC, Canada*

HAIYING HANG • *Key Laboratory for Protein and Peptide Pharmaceuticals, Institute of Biophysics, Chinese Academy of Sciences, Beijing, China*

YAXING HAO • *Institute of Immunology, Third Military Medical University, Chongqing, China*

ALEXANDER HOFFMANN • *Department of Microbiology, Immunology, and Molecular Genetics, Institute for Quantitative and Computational Biosciences, University of California, Los Angeles, CA, USA*

JIANJUN HU • *Institute of Immunology, Third Military Medical University, Chongqing, China*

QIZHAO HUANG • *Institute of Immunology, Third Military Medical University, Chongqing, China*

IL-YOUNG HWANG • *B Cell Molecular Immunology Section, Laboratory of Immunoregulation, National Institute of Allergy and Infectious Diseases, National Institutes of Health, Bethesda, MD, USA*

JOHN H. KEHRL • *Laboratory of Immunoregulation, National Institute of Allergy and Infectious Diseases, National Institutes of Health, Bethesda, MD, USA*

KATHRIN KLÄSENER • *Department of Molecular Immunology, BIOSS Centre for Biological Signaling Studies, Biology III, Faculty of Biology, University of Freiburg, Freiburg, Germany; Max Planck Institute of Immunobiology and Epigenetics, Freiburg, Germany*

ZHIRONG LI • *Institute of Immunology, Third Military Medical University, Chongqing, China*

YIDING LI • *Institute of Immunology, Third Military Medical University, Chongqing, China*

LIN LIN • *Department of Clinical Laboratory, Ruijin Hospital, Shanghai Jiaotong University School of Medicine, Shanghai, China*

WANLI LIU • *MOE Key Laboratory of Protein Sciences, School of Life Sciences (SLS), Collaborative Innovation Center for Diagnosis and Treatment of Infectious Diseases (CCID), Institute for Immunology of Tsinghua University (IITU), Tsinghua University, Beijing, China*

XIAOBING LIU • *Institute of Immunology, Third Military Medical University, Chongqing, China*

RUIQI LUO • *Key Laboratory for Protein and Peptide Pharmaceuticals, Institute of Biophysics, Chinese Academy of Sciences, Beijing, China*

PIETA K. MATTILA • *Institute of Biomedicine, Unit of Pathology, and MediCity Research Laboratories, University of Turku, Turku, Finland*

HEATHER MILLER • *Laboratory of Intracellular Parasites, Rocky Mountain Laboratories, National Institute of Allergy and Infectious Diseases, National Institutes of Health, Hamilton, MT, USA*

ANURAG R. MISHRA • *Department of Biosciences and Bioengineering, Indian Institute of Technology, Indore, India*

SIMON MITCHELL • *Department of Microbiology, Immunology, and Molecular Genetics, Institute for Quantitative and Computational Biosciences, University of California, Los Angeles, CA, USA*

CHUNG PARK • *B Cell Molecular Immunology Section, Laboratory of Immunoregulation, National Institute of Allergy and Infectious Diseases, National Institutes of Health, Bethesda, MD, USA*

MICHAEL RETH • *Department of Molecular Immunology, BIOSS Centre for Biological Signaling Studies, Biology III, Faculty of Biology, University of Freiburg, Freiburg, Germany; Max Planck Institute of Immunobiology and Epigenetics, Freiburg, Germany*

IVAN REY • *Biophysics Program, University of Maryland, College Park, MD, USA*

KOUSHIK ROY • *Department of Microbiology, Immunology, and Molecular Genetics, Institute for Quantitative and Computational Biosciences, University of California, Los Angeles, CA, USA*

MAXIM NIKOLAIEVICH SHOKHIREV • *The Salk Institute for Biological Studies, La Jolla, CA, USA*

HAE WON SOHN • *Laboratory of Immunogenetics, National Institute of Allergy and Infectious Diseases, National Institutes of Health, Rockville, MD, USA*

WENXIA SONG • *Department of Cell Biology and Molecular Genetics, University of Maryland, College Park, MD, USA*

KATELYN M. SPILLANE • *Immune Receptor Activation Laboratory, The Francis Crick Institute, London, UK; Division of Immunology and Inflammation, Department of Medicine, Imperial College London, London, UK*

VID ŠUŠTAR • *Institute of Biomedicine, Unit of Pathology, and MediCity Research Laboratories, University of Turku, Turku, Finland*

PAVEL TOLAR • *Immune Receptor Activation Laboratory, The Francis Crick Institute, London, UK; Division of Immunology and Inflammation, Department of Medicine, Imperial College London, London, UK*

BEBHINN TREANOR • *Department of Immunology, University of Toronto, Toronto, ON, Canada; Department of Cell and Systems Biology, University of Toronto, Toronto, ON, Canada; Department of Biological Sciences, University of Toronto Scarborough, Toronto, ON, Canada*

ARPITA UPADHYAYA • *Biophysics Program, University of Maryland, College Park, MD, USA; Department of Physics, University of Maryland, College Park, MD, USA; Institute for Physical Science and Technology, University of Maryland, College Park, MD, USA*

MARIKA VAINIO • *Institute of Biomedicine, Unit of Pathology, and MediCity Research Laboratories, University of Turku, Turku, Finland*

ZHENGPENG WAN • *MOE Key Laboratory of Protein Sciences, Collaborative Innovation Center for Diagnosis and Treatment of Infectious Diseases, School of Life Sciences, Tsinghua University, Beijing, People's Republic of China*

HAOQIANG WANG • *Institute of Immunology, Third Military Medical University, Chongqing, China*

JIA C. WANG • *Department of Microbiology and Immunology and the Life Sciences Institute, University of British Columbia, Vancouver, BC, Canada*

JING WANG • *MOE Key Laboratory of Protein Sciences, Collaborative Innovation Center for Diagnosis and Treatment of Infectious Diseases, School of Life Sciences, Tsinghua University, Beijing, People's Republic of China*

PENGCHENG WANG • *Institute of Immunology, Third Military Medical University, Chongqing, China*

YIFEI WANG • *The First Affiliated Hospital, Biomedical Translational Research Institute and, Guangdong Province Key Laboratory of Molecular Immunology and Antibody Engineering, Jinan University, Guangzhou, China; Institute of Immunology, Third Military Medical University, Chongqing, China*

LAABIAH WASIM • *Department of Immunology, University of Toronto, Toronto, ON, Canada*

BRITTANY A. WHEATLEY • *Department of Physics, University of Maryland, College Park, MD, USA*

LIFAN XU • *Institute of Immunology, Third Military Medical University, Chongqing, China*

JIANYING YANG • *Department of Molecular Immunology, BIOSS Centre for Biological Signaling Studies, Biology III, Faculty of Biology, University of Freiburg, Freiburg, Germany; Max Planck Institute of Immunobiology and Epigenetics, Freiburg, Germany*

ZHIYONG YANG • *Cardiovascular Research Institute, University of California, San Francisco, San Francisco, CA, USA; Sandler Asthma Basic Research Center, University of California, San Francisco, San Francisco, CA, USA*

DI YU • *China-Australia Centre for Personalised Immunology, Renji Hospital, Shanghai Jiaotong University School of Medicine, Shanghai, China; Department of Immunology and Infectious Disease, John Curtin School of Medical Research, The Australian National University, Canberra, ACT, Australia*

LILIN YE • *Institute of Immunology, Third Military Medical University, Chongqing, China*

YUN ZHAO • *Key Laboratory for Protein and Peptide Pharmaceuticals, Institute of Biophysics, Chinese Academy of Sciences, Beijing, China*

Chapter 1

Activation-Induced Cytidine Deaminase Aided In Vitro Antibody Evolution

Lili An, Chuan Chen, Ruiqi Luo, Yun Zhao, and Haiying Hang

Abstract

Activation-induced cytidine deaminase (AID) initiates somatic hypermutation (SHM) by converting deoxycytidines (dC) to deoxyuracils (dU) which then can induce other mutations, and plays a central role in introducing diversification of the antibody repertoire in B cells. Ectopic expression of AID in bacteria and non-B cells can also lead to frequent mutations in highly expressed genes. Taking advantage of this feature of AID, in recent years, systems coupling in vitro somatic hypermutation and mammalian cell surface display have been developed, with unique benefits in antibody discovery and optimization in vitro. Here, we provide a protocol for AID mediated in vitro protein evolution. A CHO cell clone bearing a single gene expression cassette has been constructed. The gene of an interested protein for in vitro evolution can be easily inserted into the cassette by dual recombinase-mediated cassette exchange (RMCE) and constantly expressed at high levels. Here, we matured an anti-TNFα antibody as an example. Firstly, we obtained a CHO cell clone highly displaying the antibody by dual RMCE. Then, the plasmid expressing AID is transfected into the CHO cells. After a few rounds of cell sorting-cell proliferation, mutant antibodies with improved features can be generated. This protocol can be applied for improving protein features based on displaying levels on cell surface and protein-protein interaction, and thus is able to enhance affinity, specificity, and stability besides others.

Key words Activation-induced cytidine deaminase (AID), Somatic hypermutation (SHM), Antibody evolution, Affinity maturation

1 Introduction

Antibodies are the antigen-binding proteins present on the B cell membrane and secreted by plasma cells. Because of the diversification of the antibody repertoire, the vertebrate immune system is able to respond to an almost limitless spectrum of foreign antigens [1]. During B cell activation, somatic hypermutation (SHM) contributes greatly to the diversification of the antibodies. SHM occurs only within germinal centers and is targeted to a rearranged variable region of antibody at a frequency approaching $10^{-4} \sim 10^{-3}$ per base pair per generation, which is several orders of magnitude higher than the "background" rate of mutation across the genome. It has

Chaohong Liu (ed.), *B Cell Receptor Signaling: Methods and Protocols*, Methods in Molecular Biology, vol. 1707,
https://doi.org/10.1007/978-1-4939-7474-0_1, © Springer Science+Business Media, LLC 2018

been demonstrated that activation-induced cytidine deaminase (AID) plays a central role in this process [2]. AID expresses specifically in germinal center B cells [3] and initiates SHM by converting deoxycytidines (dC) to deoxyuracils (dU) which are subsequently processed by one of several mechanisms and lead to both transition and transversion mutations, sometimes even deletion, during antibody affinity maturation [4, 5]. Interestingly, ectopic expression of AID in bacteria and non-B cells, such as H1299 cells and fibroblast cell lines, can also lead to high-frequency mutations on highly expressed genes [6–8]. The gene transcriptional level is positively proportional to the mutation frequency, and thus to increased chances to generate protein features sought for.

Normal serum contains a mixture of heterogeneous antibodies. These polyclonal antibodies have obvious advantage for the immune response of an organism in vivo. While monoclonal antibodies, which are specific for a single epitope, are more preferable in most researches and medical applications. They have been proven to be very useful for diagnostic, imaging, and therapeutic purpose in clinical medicine. The development of in vivo immunization-based hybridization techniques allowed for the production of monoclonal antibodies in many applications [9]. However, this technique is time-consuming and harbor-intensive, and has shortcomings in the production of the antibodies from the conserved antigens with poor immunogenicity or toxic antigens. In vitro display systems such as phage, yeast, and bacterial displays, which overcome the shortcomings of hybridization techniques, have been used in antibody discovery and optimization [10]. Because of the disadvantages in protein folding, posttranslational modification, and code usage, there are limitations in the phage, yeast, and other microbial expression-based systems for antibody evolution. In recent years, systems coupling in vitro somatic hypermutation and mammalian cell surface display have been developed by several groups including our research team [8, 11–13].

Here, we provide a protocol for AID aided in vitro antibody evolution. In this protocol, we used a CHO cell clone (CHO-puro) with a retargetable expression cassette, flanked with FRT and LoxP sequences, in only one genomic locus that harbored a puromycin resistance gene (*puroR*) able to be highly expressed. A DNA fragment of interest, also flanked with FRT and LoxP sequences, can replace *puroR* by simultaneously expressing Flp and Cre recombinases (dual RMCE: recombinase-mediated cassette exchange) [14]. We first obtained a CHO cell clone highly displaying the antibody by dual RMCE. Then, the plasmid highly expressing AID was transfected into the CHO cells. After a few rounds of cell sorting-cell proliferation, mutant antibodies with improved affinities can be generated [15].

2 Materials

2.1 Purification of GFP-TNFa from E. coli

1. The plasmid PET28a(+)-GFP-TNFα for GFP-TNFα protein expression in *E. coli* has been constructed previously [8].

2. Lysogeny broth (LB) media:10 g/l Tryptone, 5 g/l Yeast extract, and 5 g/l NaCl.

3. 50 mg/ml kanamycin: Dissolve 2.5 g kanamycin in 50 ml double distilled water (ddH$_2$O).

4. 100 mM isopropytlthiogalactoside (IPTG):1.2 g IPTG dissolved in 50 ml ddH$_2$O.

5. 100 mM Phenylemthanesulfonyl fluoride (PMSF): 0.174 g PMSF dissolved in 10 ml Isopropyl alcohol.

6. Binding buffer:500 mM NaCl, 20 mM Tris-Cl, 10% glycerol, 5 mM imidazole, pH 7.9.

7. Bradford protein assay kit (Bio-Rad).

8. Ultrasonic instrument (Nanjing Xin Chen Biological Technology Co., Ltd).

9. Ni Sepharose High Performance column (Amersham Biosciences).

10. Constant temperature oscillation incubator HZQ-×100.

2.2 Cell Culture and Transfection

1. Cells:

   (a) CHO/dhFr− cells have been purchased from Cell Bank of the Chinese Academy of Sciences, Shanghai, China.

   (b) CHO-puro cells (The CHO/dhFr− cells with only one copy of retargetable high-level expression cassette) have been established by RMCE previously and explained in detail [15].

   (c) FreeStyle™ MAX 293F cells have been purchased from Invitrogen.

2. Cell culture media:

   (a) CHO/dhFr− cells, CHO-puro cells and other cell lines derived from CHO/dhFr− cells are propagated in Iscove's Modified Dulbecco Medium (IMDM, HyClone) containing 10% fetal bovine serum (FCS, HyClone), 0.1 mM hypoxanthin, and 0.016 mM thymidine (HT, Gibco, USA).

   (b) FreeStyle™ MAX 293F cells are propagated in Free-Style™ 293 Expression Medium (Invitrogen).

3. The plasmids pF2AC for coexpression of Flpo and iCre recombinase, pFAbL for the replacement of the *puroR* with the antibody (TNFα antibody) gene by dual RMCE in a predetermined genomic locus of CHO-puro cells, pCEP-Neo-mAID

for the expression of mAID and pCEP4-scFv-hFC for the expression of scFv-hFc fusion protein in FreeStyle™ MAX 293F cells have been constructed previously and explained in detail [15].

4. Lipofectamine™ 2000 (Invitrogen) and Opti-MEM® I Reduced Serum Medium (Gibco) have been used for the transfection of CHO/dhFr− cells, CHO-puro cells, and other cell lines derived from CHO/dhFr− cells

5. Linear Polyethylenimine 25,000 (PEI) reagent (Polysciences, Inc) and OptiPRO™ SFM Serum Free Medium (Gibco) have been used for the transfection of FreeStyle™ MAX 293F cells.

6. 100 mg/ml neomycin: Dissolve 1 g neomycin in 50 ml sterile ddH$_2$O.

7. The Steri-Cycle CO$_2$ Incubator (Thermo) has been used for the culture of CHO/dhFr− cells, CHO-puro cells, and other cell lines derived from CHO/dhFr− cells.

8. The Shaker-Incubator Lab-Therm/LT-XC (Kuhner, Switzerland) has been used for the culture of FreeStyle™ MAX 293F cells.

2.3 Flow Cytometric Detection and Soring

1. Anti-HA-PE antibody (Abcam) has been used for cell labeling.
2. FACSCalibur (BD).
3. FACSAria™IIIu (BD).

2.4 Detection of Antibody Mutations

1. Wizard® SV Genomic DNA Purification System (Promega) has been used for cell genomic DNA extraction.
2. Phusion High-Fidelity PCR Kit (NEB).
3. AxyPrep™ DNA Gel Extraction Kit (Axygen) has been used for nucleic acid purification.
4. EcoRI and XhoI.
5. T4 DNA ligase (Thermo Scientific).
6. S1000™ Thermal Cyclers (Bio-Rad).
7. CHB-100 thermostatic metal bath (Bioer).
8. The Nanodrop ND-1000 Spectrophotometer (Thermo Scientific).

2.5 Octet Biomolecular Interaction Technology

1. Octet biomolecular interaction technology Platform (ForteBio Octet, USA).
2. Anti-Fc (human Fc) kinetic grade biosensors (ForteBio:18-5060, USA).

3   Methods

In this protocol, we first obtain the CHO cells highly displaying the antibody by dual RMCE. Then, the plasmid expressing AID are transfected into the CHO cells. After a few rounds of flow sorting-cell proliferation, a mutant antibody with improved affinity was generated [15].

3.1 Preparation and Purification of GFP-TNFα

As an example, GFP-TNFα fusion protein has been used as the antigen for TNFα antibody screening (*see* **Note 1**).

1. Transfer the Rosetta cells harboring His-GFP-TNFα expression plasmid from the single colony into the tube with 5 ml LB medium containing 50 μg/ml kanamycin and incubate the tube at 37 °C overnight with shaking at 200 rpm.

2. Transfer 2 ml of the *E. coli* Rosetta cell cultures into 200 ml fresh LB medium containing 50 μg/ml kanamycin and incubate at 37 °C with shaking at 200 rpm until the culture density reaches 0.5~0.6 OD600.

3. Add IPTG to a final concentration of 0.4 mM and the cells are grown at 16 °C for 11 h.

4. Harvest the cells by centrifugation at $2600 \times g$ for 10 min and discard the supernatant.

5. Resuspend the cell pellet in 10 ml PBS containing 2 mM PMSF at 4 °C.

6. Sonication: Amplitude, 25%; Pulse on/off, 5 s/10 s; 30 min.

7. Centrifugate at $16,000 \times g$ for 30 min at 4 °C.

8. Add the supernatant to a 50% slurry of Ni Sepharose High Performance column.

9. Rinse the column with 30, 60, 90, and 250 mM imidazole in binding buffer in a stepwise manner.

10. After purification, measure the concentration of these recombinant proteins with Bradford protein assay kit (Bio-Rad) according to the standard procedure.

In most cases, the purity of GFP-TNFα is more than 90%.

3.2 Construction of the Cells Stably Displaying the Antibody

The CHO cell clone (CHO-puro) with a retargetable expression cassette in only one genomic locus that harbored a puromycin resistance gene (*puroR*) able to be highly expressed has been established previously [15]. This *puroR* can be replaced by different DNA fragments of interest. As an example, we replaced the *puroR* with TNFα antibody gene to obtain the CHO cells stably displaying the TNFα antibody for evolution (*see* **Note 2**).

| 3.2.1 *Cell Culture* | 1. Propagate CHO-puro cells in the culture medium at 37 °C in a 5% CO_2 incubator. |

1. Propagate CHO-puro cells in the culture medium at 37 °C in a 5% CO_2 incubator.

2. Subculture cells every 2~3 days.

3. One day before transfection, plate 2×10^6 cells in 10 ml of culture medium in a 10 cm dish.

3.2.2 Transfection

To obtain TNFα antibody expressing cells (named CHO-TNFAb), CHO-puro cells are co-transfected with pFAbL (for the replacement of the *puroR* with TNFα antibody gene) as well as pF2AC (for coexpression of Flpo and iCre recombinase), and expanded for 2 days. For each transfection sample, prepare complexes as follows:

1. Dilute 16 μg pF2AC and 4 μg pFAbL in 1.2 ml of Opti-MEM® I Reduced Serum Medium without serum. Mix gently.

2. Dilute 40 μl Lipofectamine™ 2000 in 1.2 ml of Opti-MEM® I Reduced Serum Medium. Incubate for 5 min at room temperature.

3. After incubation for 5 min, combine the diluted DNA with diluted Lipofectamine™ 2000. Mix gently and incubate for 25 min at room temperature.

4. Discard the medium from the CHO-puro cell culture plate. Wash CHO-puro cells with PBS for three times. After washing, add 4 ml of the Opti-MEM® I Reduced Serum Medium into the plate.

5. Add 2.4 ml of complexes (from **step 3**) to each dish containing cells and medium. Mix gently by rocking the dish back and forth.

6. Incubate cells at 37 °C in a CO_2 incubator for 6 h.

3.2.3 Cell Expansion

1. Six hours after transfection, split the cells into four 10 cm dishes.

2. Expand the cells for 2 days.

3.2.4 Flow Cytometric Sorting

1. Detach cells by adding the trypsin with room temperature to the culture dishes (2 ml trypsin to a 10 cm dish), incubate for 2~3 min at 37 °C (add the medium containing 10% serum to the dish right after the cells are detached) (*see* **Note 3**), and harvest the cells.

2. Resuspend the cell pellet with PBS and centrifuge at $300 \times g$ for 3 min. Aspirate the PBS.

3. The cells are incubated with 1 ml Opti-MEM® I Reduced Serum Medium containing PE-conjugated anti-HA antibody (1:250) for 25 min at 4 °C.

4. Resuspend the cell pellet with PBS and centrifuge at $300 \times g$ for 3 min. Aspirate the PBS. Repeat this process for three times.

5. Resuspend the cells with 1 ml Opti-MEM® I Reduced Serum Medium for flow sorting.

6. Sort the cells with the highest PE fluorescence on a FACSAria™IIIu.

3.2.5 Expansion of the Sorted Cells

The sorted CHO-TNFAb cells are propagated in the culture medium at 37 °C in a 5% CO_2 incubator.

3.3 Affinity Maturation of Anti-TNFα Antibody

Antibody affinity maturation is carried out with a few rounds of cell proliferation-flow sorting as described below.

3.3.1 Transfection

For each round of maturation, CHO-TNFAb cells are transfected with the AID-expressing plasmid pCEP-Neo-mAID.

1. One day before transfection, plate 3×10^5 CHO-TNFAb cells in 2 ml of the culture medium in a 6-well plate.

2. Dilute 2 μg pCEP-Neo-mAID in 150 μl of Opti-MEM® I Reduced Serum Medium without serum. Mix gently.

3. Dilute 5 μl Lipofectamine™ 2000 in 150 μl of Opti-MEM® I Medium. Incubate for 5 min at room temperature.

4. After incubation for 5 min, combine the diluted DNA with diluted Lipofectamine™ 2000. Mix gently and incubate for 25 min at room temperature.

5. Discard the medium from the CHO-TNFAb cell culture plate. Wash CHO-TNFAb cells with PBS for three times. After washing, add 1 ml of the Opti-MEM® I Reduced Serum Medium into the well.

6. Add the 300 μl of complexes to each well containing cells and medium. Mix gently by rocking the plate back and forth.

7. Incubate the cells at 37 °C in a CO_2 incubator for 5 h.

3.3.2 Cell Expansion

Expand AID expression cells for the accumulation of antibody gene mutations in the neomycin-containing medium (to keep the AID-expressing plasmid in cells) for 7 days.

1. Five hours after transfection, passage the cells from the 6-well plate into two 10 cm dishes.

2. Add neomycin to the final concentration of 1 mg/ml to keep the AID-expressing plasmid in cells.

3. Expand cells for 7 days with subcultures every 2~3 days until the number of the cells reached about 6×10^7.

3.3.3 Enrichment of Cells Displaying High-Affinity Antibodies

The cells are labeled with PE-conjugated anti-HA antibody (to monitor antibody display levels on cells), and GFP-TNFα antigen (to monitor the TNF-α-binding ability displayed on the cell

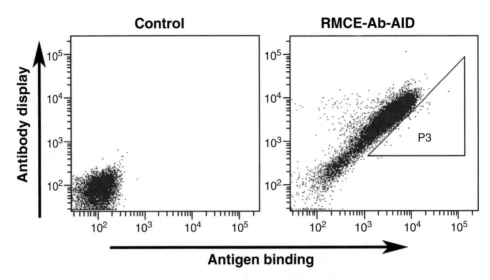

Fig. 1 Antibody affinity maturation. The cells with the highest "antigen binding/antibody display ratio" (the top 0.02%, around 20,000 cells) are collected by flow sorting. This figure was from ref. 15 (Chen, C., et al. *Biotechnol Bioeng* 2016, **113**, 39–51)

surface), and the labeled cells with the highest GFP/PE fluorescence ratio (the top 0.02%, around 20,000 cells) are collected by flow sorting which are detailed below.

1. Detach the cells by adding the trypsin with room temperature to the culture dishes (2 ml trypsin to a 10 cm dish), incubate for 2~3 min at 37 °C (add the medium containing 10% serum to the dish right after the cells are detached) (*see* **Note 3**), and harvest the cells.

2. Resuspend the cell pellet with PBS and centrifuge at $300 \times g$ for 3 min. Aspirate the PBS.

3. The cells are incubated with 2 ml Opti-MEM® I Reduced Serum Medium containing PE-conjugated anti-HA antibody (1:250) and GFP-TNFα (0.2 mg/ml) for 25 min at 4 °C.

4. Resuspend the cell pellet with PBS and centrifuge at $800 \times g$ for 3 min. Aspirate the PBS. Repeat this process for three times.

5. Resuspend the cells with 2 ml Opti-MEM® I Reduced Serum Medium for FACS.

6. The labeled cells with the highest antigen-binding/antibody display ratio (the top 0.02%, around 20,000 cells) are collected by flow sorting (Fig. 1) [15].

3.3.4 Expansion of the Sorted Cells

The sorted CHO-TNFAb cells are propagated in the culture medium at 37 °C in a 5% CO_2 incubator. The collected cells are grown for 3 days, and the cell number will reach around 200,000. The cells are then transfected with the AID-expressing plasmid for the next round of maturation.

3.4 Detection of the Mutations

1. Harvest the cells.

2. Extracted genomic DNA with Wizard® SV Genomic DNA Purification System according to the manufacturer's instruction. The DNA concentration is determined with NanoDrop and adjusted to 100 ng/μl.

3. Amplify antibody gene fragments from genomic DNA. Use the following primers for antibody gene PCR:

   T7-P1: 5′-TAATACGACTCACTATAGGGAGACCCAAGC-3′

   TM-P2: 5′-CTAACGTGGCTTCTTCTGCCAAAGC-3′

   Briefly vortex and centrifuge all reagents before setting up the reactions.

   For each 50 μl reaction, add the following components to a PCR tube:

   10 μl 5× HF buffer.

   1 μl 10 mM dNTPs.

   1.5 μl DMSO.

   0.5 μl Phusion® High-Fidelity DNA Polymerase.

   2.0 μl forward primer (10 μM).

   2.0 μl reverse primer (10 μM).

   29 μl Nuclease-free water.

   4 μl DNA template (100 ng/μl).

   Run the following PCR protocol:
   Initial denaturation: 98 °C, 3 min.

   30 cycles of 98 °C, 30 s; 61 °C, 30 s; 72 °C, 35 s.

   Final elongation: 72 °C, 10 min.

   Hold at 4 °C.

4. Purification of the 1077 bp PCR fragment using AxyPrep™ DNA Gel Extraction Kit.

5. Digest PCR fragment and pCDNA3.1 plasmid with EcoRI and XhoI at 37 °C for 2 h, then purify with AxyPrep™ DNA Gel Extraction Kit.

6. Ligate the digested PCR fragment with the linearized plasmid vector using T4 DNA ligase at 16 °C for 4 h.

7. Transform chemically competent *E. coli* with ligation products. After each round of sorting, the antibody genes are cloned, and 50 clones are sequenced (*see* **Note 4**). As cell expansion-cell sorting is iterated, beneficial mutations for higher affinity will be accumulated (*see* **Note 5**).

3.5 Detection of Antibody Affinity Using Flow Cytometry

All the antibody genes containing unique mutations or mutation combinations are analyzed with flow cytometry for their abilities to bind GFP-TNFα (*see* **Note 6**).

3.5.1 Cell Culture	1. Propagate CHO/dhFr− cells in the culture medium at 37 °C in a 5% CO_2 incubator.
	2. Subculture cells every 2–3 days.
	3. One day before transfection, plate 2×10^5 cells in 2 ml of the culture medium in a 6-well plate.

3.5.2 Transfection

The pCDNA3.1 plasmids harboring the antibody genes containing unique mutations or mutation combinations are transfected into CHO/dhFr− cells. For each transfection sample, prepare complexes as follows:

1. Dilute 2 μg plasmid DNA in 150 μl of Opti-MEM® I Reduced Serum Medium without serum. Mix gently.

2. Mix Lipofectamine™ 2000 gently before use, then dilute 5 μl Lipofectamine™ 2000 in 150 μl of Opti-MEM® I Reduced Serum Medium. Incubate for 5 min at room temperature.

3. After the 5 min incubation, combine the diluted DNA with diluted Lipofectamine™ 2000. Mix gently and incubate for 25 min at room temperature.

4. Discard the medium from the CHO-TNFAb cell culture plate. Wash CHO-TNFAb cells with PBS for three times. After washing, add 1 ml of the Opti-MEM® I Reduced Serum Medium into the well.

5. Add the 300 μl of complexes to each dish containing cells and medium. Mix gently by rocking the dish back and forth.

6. Incubate the cells at 37 °C in a CO_2 incubator for 5 h. Substitute the medium with the culture medium.

3.5.3 Flow Cytometry Detection

1. Two days after transfection, detach cells by adding the trypsin with room temperature to the culture dishes (0.5 ml trypsin to a well), incubate for 2~3 min at 37 °C (add the medium containing 10% serum to the dish right after the cells are detached) (*see* **Note 3**), and harvest the cells.

2. Resuspend the cell pellet with PBS and centrifuge at $300 \times g$ for 3 min. Aspirate the PBS.

3. The cells are incubated with 100 μl Opti-MEM® I Reduced Serum Medium containing PE-conjugated anti-HA antibody (1:250) and GFP-TNFα (0.2 mg/ml) for 25 min at 4 °C.

4. Resuspend the cell pellet with PBS and centrifuge at $300 \times g$ for 3 min. Aspirate the PBS. Repeat this process for three times.

5. Resuspend the cells with 0.4 ml Opti-MEM® I Reduced Serum Medium for flow cytometry detection.

As an example the mutant antibodies demonstrated stronger capabilities to bind GFP-TNFα than that of the wild-type antibody. That is, the "antigen binding/antibody display ratio" of the cells

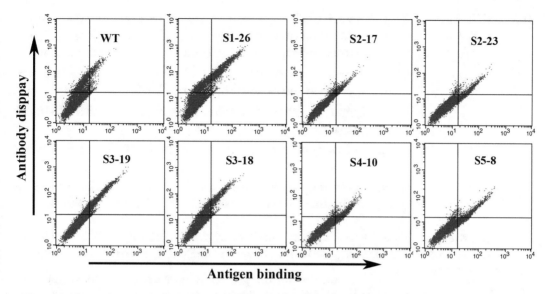

Fig. 2 Detection of the levels of antibody display and antigen-binding. The levels of antibody display and antigen-binding were examined on cells transiently transfected with designated mutant antibody genes, respectively. This figure was from ref. 15 (Chen, C., et al. *Biotechnol Bioeng 2016, **113**, 39–51*)

displaying these antibodies was larger than that of the parent antibody (Fig. 2) [15].

3.6 Detection of Antibody Affinity by Octet Biomolecular Interaction Technology

As the example, the genes of seven most frequently appeared antibody mutants were subcloned into pCEP4 plasmids and transfected into 293F cells to produce scFv-Fc fusion protein for Octet biomolecular interaction detection.

3.6.1 Cell Culture

1. Propagate FreeStyle™ MAX 293F cells in FreeStyle™ 293 Expression Medium, at 37 °C in a 5% CO_2 incubator with shaking at 120 rpm.

2. Subculture cells every 2~3 days.

3. One day before transfection, inoculate 293F cells in 30 ml FreeStyle™ 293 Expression Medium at the concentration of 8×10^5 cells/ml, so that the cells will reach 1.2×10^6 cells/ml at the time of transfection.

3.6.2 Transfection

For each transfection sample, prepare complexes as follows:

1. Dilute 45 µg of plasmid DNA in OptiPRO™ SFM Serum Free Medium to a total volume of 2 ml. Mix gently.

2. Add 60 µl of PEI reagent in **step 1**. Mix gently and incubate for 10 min at room temperature.

3. Add **step 2** to the flask containing 293F cells for transfection.

4. Incubate the cells at 37 °C in a 5% CO_2 incubator with shaking at 120 rpm.

3.6.3 Production and Purification of scFv-Fc

1. Three days after transfection, transfer the cells and the medium into a 50 ml centrifuge tube. Centrifuge at $300 \times g$ for 3 min. Transfer the supernatant into a new 50 ml centrifuge tube.

2. Centrifuge at $7000 \times g$ for 10 min at 4 °C. Transfer the supernatant into a new 50 ml centrifuge tube.

3. Filtrate the supernatant using the syringe filter with a 0.45 μm pore size.

4. Concentrate the supernatant from about 30 ml to 1.5 ml using 30 KDa concentrator by centrifuging at $1500 \times g$ at 4 °C.

5. Adjust the concentration of the scFv-Fc fusion protein to 100 μg/ml using the FreeStyle™ 293 Expression Medium.

3.6.4 Detection of Antibody Affinity by Octet Biomolecular Interaction Technology

Antibody affinity is measured by the Octet biomolecular interaction technology according to the manufacturer's instruction.

1. The scFv-Fc fusion protein supernatants is concentrated to 100 mg/ml, then detected by seven Anti-Fc kinetic grade biosensors.

2. The detection conditions used are (a) baseline 240 s; (b) loading 240 s; (c) baseline 180 s; (d) association 120 s with a series of concentrations (1600 nM, 800 nM, 400 nM, 200 nM, 100 nM, 50 nM, 25 nM) of GFP-TNFα antigen; (e) dissociation 180 s.

3. The K_{on} and K_{off} rates are measured by Octet software and K_d is calculated for each antibody mutation by the K_{off}/K_{on} ratio.

As an example, seven most frequently appeared antibody mutants as shown in Fig. 2 were further analyzed by Octet biomolecular interaction technology for their dissociation coefficients (K_d), confirming their improved affinities (Table 1) [15].

Table 1
Affinities of different mutations

Clone	Mutation	K_d (M)	K_{off} (1/s)	K_{on} (1/MS)
WT	No	7.82E−09	5.29E−04	6.76E+04
S1-26	A229T	5.69E−09	4.07E−04	7.15E+04
S2-17	N33K	4.34E−09	2.58E−04	5.96E+04
S3-19	A145V	2.12E−09	1.57E−04	7.40E+04
S2-23	N33K+A145V	1.25E−09	9.78E−05	7.81E+04
S3-18	A145V+M211I	1.32E−09	1.04E−04	7.87E+04
S4-10	N33K+H212N	2.88E−09	1.66E−04	5.77E+04
S5-8	N33K+A145V+ M211I	5.20E−10	3.94E−05	7.84E+04

Note: This table was from ref. 15 (Chen, C., et al. *Biotechnol Bioeng* 2016, *113*, 39–51)

4   Notes

1. As shown in this example, GFP-TNFα fusion protein purified from *E. coli*. harboring GFP-TNFα expression plasmid, in which the TNFα coding sequence was directly fused to the sequence encoding the C terminus of GFP without linker sequence, was used as the antigen for TNFα antibody screening. This fusion protein has been proven works well [8, 15]. As to different antigen-antibody interactions, the location of the tag and the suitable linker between the tag or the reporter molecule and the antigen should be taken into consideration to ensure the correct structure of the antigen.

2. As to the cells stably displaying the antibody, the following two points should be noticed. (1) To couple the antibody phenotype to genotype, the integrated antibody gene for evolution should be one copy per cell; (2) Since the gene transcriptional level is positively proportional to the mutation frequency, the antibody gene for evolution should be highly transcribed.

3. Care has been taken to detach the CHO cells with trypsin digestion. Some displayed antibodies are sensitive to this process. It is necessary to do the preliminary experiments to test the effects of the time of trypsin digestion. As for antibodies or other proteins extremely sensitive to trypsin digestion, we recommend detaching the CHO cells by incubating 0.02% EDTA at 37 °C for 5 min.

4. As to the first two rounds of sorting, it is recommended that 100 or 150 clones are sequenced.

5. As cell expansion-cell sorting is iterated, beneficial mutations for higher affinity will be accumulated. As an example, five rounds of cell expansion and cell sorting enriched mutant antibody clones with improved affinities; the clone with mutations of N33K/A145V/M211I has an affinity 15 times higher than the parent clone (Table 1) [15]. In our hands, much faster affinity maturations were achieved with more stringent sorting or more active AID enzymes (data not published).

6. As shown in this example, we obtained the cells with high antigen affinity and got the mutants of the antibody genes from these cells. However, AID can act outside of the antibody gene. Thus, the increased affinity of the cells in the sorting window might be because of the changes of the cells, not the mutations of the antibody genes. To exclude this, it is necessary to transiently transfect the antibody genes containing unique mutations or mutation combinations into the wild-type CHO/dhFr− cells to confirm their abilities to bind the antigen.

Acknowledgments

This work was supported by the grant from Ministry of Science and Technology of the People's Republic of China (grant number 2014CB910402) and the National Natural Science Foundation of China (grant number 31370792).

References

1. Fear DJ (2013) Mechanisms regulating the targeting and activity of activation induced cytidine deaminase. Curr Opin Immunol 25:619–628

2. Teng G, Papavasiliou FN (2007) Immunoglobulin somatic hypermutation. Annu Rev Genet 41:107–120

3. Muramatsu M, Sankaranand VS, Anant S, Sugai M, Kinoshita K, Davidson NO, Honjo T (1999) Specific expression of activation-induced cytidine deaminase (AID), a novel member of the RNA-editing deaminase family in germinal center B cells. J Biol Chem 274:18470–18476

4. Odegard VH, Schatz DG (2006) Targeting of somatic hypermutation. Nat Rev Immunol 6:573–583

5. Stavnezer J (2011) Complex regulation and function of activation-induced cytidine deaminase. Trends Immunol 32:194–201

6. Petersenmahrt SK, Harris RS, Neuberger MS (2002) AID mutates E. coli suggesting a DNA deamination mechanism for antibody diversification. Nature 418:99–103

7. Yoshikawa K, Okazaki IM, Eto T, Kinoshita K, Muramatsu M, Nagaoka H, Honjo T (2002) AID enzyme-induced hypermutation in an actively transcribed gene in fibroblasts. Science 296:2033–2036

8. Shaopeng C, Junkang Q, Cuan C, Chunchun L, Yuheng L, Lili A, Junying J, Jie T, Lijun W, Haiying H (2012) Affinity maturation of anti-TNF-alpha scFv with somatic hypermutation in non-B cells. Protein Cell 3:460–469

9. Köhler G, Milstein C (1975) Continuous cultures of fused cells secreting antibody of predefined specificity. Nature 256:495–497

10. Bradbury AR, Sidhu S, Dübel S, McCafferty J (2011) Beyond natural antibodies: the power of in vitro display technologies. Nat Biotechnol 29:245–254

11. Arakawa H, Kudo H, Batrak V, Caldwell RB, Rieger MA, Ellwart JW, Buerstedde JM (2008) Protein evolution by hypermutation and selection in the B cell line DT40. Nucleic Acids Res 36:e1

12. Bowers PM, Horlick RA, Kehry MR, Neben TY, Tomlinson GL, Altobell L, Zhang X, Macomber JL, Krapf IP, Wu BF, McConnell AD, Chau B, Berkebile AD, Hare E, Verdino P, King DJ (2014) Mammalian cell display for the discovery and optimization of antibody therapeutics. Methods 65:44–56

13. McConnell AD, Do M, Neben TY, Spasojevic V, MacLaren J, Chen AP, Altobell L III, Macomber JL, Berkebile AD, Horlick RA, Bowers PM, King DJ (2012) High affinity humanized antibodies without making hybridomas; immunization paired with mammalian cell display and in vitro somatic hypermutation. PLoS One 7:e49458

14. Anderson RP, Voziyanova E, Voziyanov Y (2012) Flp and Cre expressed from Flp-2A-Cre and Flp-IRES-Cre transcription units mediate the highest level of dual recombinase-mediated cassette exchange. Nucleic Acids Res 40:e62

15. Chen C, Li N, Zhao Y, Hang H (2016) Coupling recombinase-mediated cassette exchange with somatic hypermutation for antibody affinity maturation in CHO cells. Biotechnol Bioeng 113:39–51

Chapter 2

Analyzing Mouse B Cell Responses Specific to LCMV Infection

Yaxing Hao, Zhirong Li, Yifei Wang, Xiaobing Liu, and Lilin Ye

Abstract

B cell responses play a central role in humoral immunity, which protects an individual from invading pathogens by antigen-specific antibodies. Understanding the basic principles of the B cell responses during viral infection is of substantial importance for anti-viral vaccine development. In inbred mice, lymphocytic choriomeningitis virus (LCMV) infection elicits robust and typical T cell-dependent B cell responses, including germinal center reaction, memory B cell formation, and a long-lived plasma cell pool in bone marrow. Therefore, this system represents an ideal model to investigate anti-viral B cell responses. In this protocol, we describe how to propagate and quantify LCMV and successfully establish an acute LCMV infection in mice. This protocol also provides three different techniques to analyze B cell responses specific to an acute LCMV infection: the identification of germinal center (GC) B cells and follicular helper CD4 T (T_{FH}) cells from the spleens and lymph nodes via flow cytometry, titration of LCMV-specific IgG in the serum after LCMV infection using an enzyme-linked immunosorbent assay (ELISA) analysis, and detection of LCMV-IgG secreted plasma cells from bone marrow with an enzyme-linked immunospot (ELISPOT) assay.

Key words LCMV, Germinal center, T_{FH} cells, Tetramer, Plasma cells, Antibody, Flow cytometry, ELISA, ELISPOT

1 Introduction

During viral infection, B cell-mediated antibody responses provide long-term protection against invading pathogens. Understanding the B cell responses during viral infection is of substantial importance for the development of improved vaccine strategies because most currently licensed human vaccines are protective T cell-dependent antibody based [1, 2].

IgM^+IgD^+ naive B cells within the follicle are activated in response to T cell-dependent antigens and migrate to the T-B border, where B cells proliferate and interact with antigen-specific follicular helper CD4 T (T_{FH}) cells, which facilitates the complete activation of B cells [3–5]. Subsequently, some activated B cells differentiate into short-lived plasmablasts, which secrete low

Chaohong Liu (ed.), *B Cell Receptor Signaling: Methods and Protocols*, Methods in Molecular Biology, vol. 1707, https://doi.org/10.1007/978-1-4939-7474-0_2, © Springer Science+Business Media, LLC 2018

affinity antibodies [6]. The other subset of activated B cells enter into the center of follicles to form the early germinal centers (GCs) [7, 8] and undergo rapid clonal expansion and somatic hypermutation (SHM) [9]. Combined with following affinity-based positive selection mediated by T_{FH} cells in the light zone of GC structures, GC B cells ultimately differentiate into long-lived antibody-secreting plasma cells and memory B cells [10]. Thus, the consequence of a germinal center reaction is to establish a high-affinity, long-term B cell memory immunity by generating memory B cells and plasma cells, which secrete high-affinity antibodies that effectively neutralize the invading pathogens [2, 11].

As an essential requirement of B cell responses, T_{FH} cells are characterized by the expression of high levels of master regulator transcription factor B cell lymphoma 6 protein (Bcl6) [12–14] and C-X-C chemokine receptor type 5 (CXCR5) [15–17], which enable T_{FH} cells to infiltrate B cell follicles and interact with cognate B cells to initiate and maintain the GC reaction. Other classic molecules associated with T_{FH} cells include programmed cell death protein 1 (PD-1), SH2 domain-containing protein 1A (SAP/SH2D1A), inducible T cell costimulator (ICOS), and interleukin 21 (IL-21) [18].

Lymphocytic choriomeningitis virus (LCMV) is a single-stranded RNA virus of the Arenaviridae family and was isolated by Charles Armstrong in St. Louis in 1933 [19]. LCMV has been used as an experimental model to understand immunological issues, including MHC restriction, T cell immunity, persistent infections, and T cell exhaustion [20, 21]. There are several identified LCMV variants, which lead to different infection outcomes based on genetic variations, virus titers, and infection routes. The Armstrong strain induces a typical acute viral infection: infection of mice with Armstrong triggers vigorous cytotoxic T lymphocyte (CTL) response that results in the complete elimination of the virus within 1–2 weeks and the establishment of $CD8^+$ T cell memory immunity [22, 23]. Moreover, virus-specific $CD4^+$ T cells are activated and differentiate into T follicular helper (T_{FH}) or type 1 helper (T_{H1}) cells in LCMV Armstrong infection. The robust T_{FH} activation contributes to strong B cell responses and the formation of the long-term antibody responses.

The LCMV infection model system provides several well-established methods and platforms to facilitate the analysis of B cell responses. For example, it is convenient to detect LCMV-specific IgG and antibody-secreted plasma cells by precoating ELISA and ELISPOT plate with LCMV-lysates; moreover, antigen-specific T_{FH} cell responses may be analyzed by introducing I-A^bgp66-77 tetramer in FACS staining. In this protocol, we describe how to propagate and quantify LCMV and more successfully establish an acute LCMV infection in mice. This protocol also provides three different techniques to analyze B cell responses specific to LCMV acute infection: the quantification of GC B cells and T_{FH} cells from the spleens and lymph nodes via flow cytometry, the titration of LCMV-specific IgG in the serum after LCMV infection using an enzyme-linked immunosorbent assay (ELISA) analysis, and the detection of LCMV-IgG secreted plasma cells from bone marrow with an enzyme-linked immunospot (ELISPOT) assay. Together with the mouse model of acute LCMV infection, these techniques will enable investigators to efficiently and comprehensively analyze virus-specific B cell responses.

2   Materials

2.1   LCMV Stock Preparation

1. Baby hamster kidney 21 cell lines (BHK-21, ATCC number: CCL-10).

2. Vero E6 cells (ATCC number: CRL-1586).

3. Fetal calf serum (FCS): Store at −20 °C.

4. Complete DMEM medium: Dulbecco's minimal essential medium (DMEM) basic, 10% FCS, 2 mM L-glutamine, 100 IU/ml penicillin, 100 μg/ml streptomycin, 55 μM beta-mercaptoethanol. Store at 4 °C.

5. DMEM 2%: DMEM, 2% FCS.

6. 2× Complete medium 199: 2× Medium 199, 20% FBS, 4 mM glutamine, 200 IU/ml penicillin, 200 μg/ml streptomycin. Store at 4 °C.

7. 1% agarose: Dissolve agarose in deionized water. Sterilize by autoclaving and store at 4 °C.

8. 1% neutral red: Dissolve neutral red in deionized water. Autoclave the solution with a stirring magnetic. Subsequently, stir the solution on a magnetic field overnight and store for up to several months at room temperature.

9. Lymphocytic choriomeningitis virus (LCMV): MOI of 0.03–0.1 pfu/cell for inoculation.

10. 75 cm^2 (T-75) or 175 cm^2 (T-175) tissue culture flasks.

11. 6-well petri plates.

12. 96-well round-bottomed plates.

13. 15 and 50 ml conical centrifuge tubes.

14. Cell scraper.

15. Water bath.

16. Sonicator (Qsonica Q500-200).

2.2   LCMV Acute Infection of Mice

1. Wild-type C57BL/6J mice aged 6–10 weeks.

2. Stock of LCMV Armstrong strain (4×10^6 pfu/ml, prepared in Subheading 3.1).

3. Dulbecco's minimal essential medium (DMEM) basic.

4. 50 ml conical centrifuge tubes.

5. 1 ml syringes.

6. Bleach solution.

7. 75% ethanol.

2.3 FACS Staining of GC B Cells and Plasma Cells

1. LCMV infected C57BL/6 mice, 8, 10, 15, and 30 days post infection.

2. Red blood cell (RBC) lysis buffer: 155 mM NH_4Cl, 12 mM $KHCO_3$, and 0.1 mM EDTA. Prepare in deionized water and adjust pH to 7.3. Store at room temperature.

3. Roswell Park Memorial Institute (RPMI) medium-1640 basic.

4. Fetal calf serum (FCS): Store at $-20\ °C$.

5. RPMI 2%: RPMI medium, 2% FCS.

6. Phosphate-buffered saline (PBS): 137 mM NaCl, 2.7 mM KCl, 10 mM Na_2HPO_4, 1.8 mM KH_2PO_4, pH 7.3. Store at room temperature.

7. Sodium azide (NaN_3): 1% (w/v) in deionized water. Store at $4\ °C$.

8. FACS staining buffer: PBS, 2% FCS, 0.01% NaN_3. Store at $4\ °C$.

9. 2% paraformaldehyde (PFA): 10 ml 16% formaldehyde solution (Thermo Scientific), 70 ml PBS. Store at $4\ °C$.

10. Anti-CD16/32 (Fc-blocker, clone 93). Store at $4\ °C$.

11. Fluorescent antibodies: FITC-Peanut Agglutinin (PNA, Vector Laboratories), PE-anti-CD95 (FAS, clone Jo2), PerCP-Cy5.5-anti-CD19 (clone eBio1D3), PE-Cy7-anti-B220 (CD45R, clone RA3-6B2), APC-anti-CD138 (clone 281-2), Fixable Viability Dye eFluor® 780 (eBioscience), V450-anti-IgD (clone 11-26c.2a). Store at $4\ °C$.

12. Dissecting instruments: Tweezers, scissors, gauze pads.

13. 60 mm × 15 mm cell culture dishes.

14. 70 μm cell strainers (BD Falcon).

15. 1 ml syringes.

16. Microscope frosted glass slides.

17. 15 ml conical tubes.

18. 96-well round-bottom plates.

19. 8 or 12 channel multichannel pipettes.

20. 5 ml FACS tubes.

21. Vortex.

22. Horizontal centrifuge.

23. Flow cytometer instrument (BD Canto II).

24. Analysis software (FlowJo, Treestar).

2.4 FACS Staining of LCMV-Glycoprotein Specific I-A^{b}gp66-77 Tetramer T$_{FH}$ Cells

1. LCMV infected C57BL/6 mice, 8–10 days post infection.

2. Red blood cell (RBC) lysis buffer: 155 mM NH$_4$Cl, 12 mM KHCO$_3$, and 0.1 mM EDTA. Dilute in deionized water and adjust pH to 7.3. Store at room temperature.

3. Roswell Park Memorial Institute (RPMI) medium-1640 basic.

4. Fetal calf serum (FCS): Store at −20 °C.

5. RPMI 2%: RPMI medium, 2% FCS.

6. Complete RPMI medium: RPMI-1640, 10% FCS, 2 mM L-glutamine, 100 IU/ml penicillin, 100 μg/ml streptomycin, 55 μM beta-mercaptoethanol. Store at 4 °C.

7. Phosphate-buffered saline (PBS): 137 mM NaCl, 2.7 mM KCl, 10 mM Na$_2$HPO$_4$, 1.8 mM KH$_2$PO$_4$, pH 7.3. Store at room temperature.

8. Sodium azide (NaN$_3$): 1% (w/v) in deionized water. Store at 4 °C.

9. FACS staining buffer: PBS, 2% FCS, 0.01% NaN$_3$. Store at 4 °C.

10. T$_{FH}$ staining buffer: PBS, 2% FCS, 0.01% NaN$_3$, 1% BSA, 2% normal mouse serum. Store at 4 °C.

11. 2% paraformaldehyde (PFA): 10 ml 16% formaldehyde solution (Thermo Scientific), 70 ml PBS. Store at 4 °C.

12. I-A^{b}gp66-77 tetramer conjugated PE or APC (NIH Tetramer Core Facility).

13. Primary antibody: Purified Rat anti-Mouse CXCR5 (BD, clone 2G8).

14. Secondary antibody: Biotin conjugated AffiniPure Goat anti-rat IgG(H+L) (Jackson ImmunoResearch).

15. Tertiary antibody: BV421-Streptavidin.

16. Other surface antibodies: Alexa488-anti-CD150 (SLAM, clone TC15-12F12.2), FITC-anti-PD-1 (CD279, clone RMP1-30), PE-anti-CD278 (ICOS, clone 7E.17G9), PerCP-Cy5.5-anti-CD4 (clone RM4-5), PE-Cy7-anti-CD44 (IM7), APC-Cy7-anti-B220 (CD45R, clone RA3-6B2), Fixable Viability Dye eFluor® 780 (eBioscience).

17. Dissecting instruments: Tweezers, scissors, gauze pads.

18. 60 mm × 15 mm cell culture dishes.

19. 70 μm cell strainers (BD Falcon).

20. 1 ml syringes.

21. Microscope frosted glass slides.

22. 15 ml conical tubes.

23. 96-well round-bottom plates.

24. 8 or 12 channel multichannel pipettes.

25. 5 ml FACS tubes.

26. Vortex.

27. Horizontal centrifuge.

28. Flow cytometer instrument (BD Canto II).

29. Analysis software (FlowJo, Treestar).

2.5 ELISA Analysis of LCMV–IgG

1. BHK-21-LCMV Clone-13 lysate: Prepared in Subheading 3.1.

2. Phosphate-buffered saline (PBS): 137 mM NaCl, 2.7 mM KCl, 10 mM Na_2HPO_4, 1.8 mM KH_2PO_4, pH 7.3. Sterilize by autoclaving and store at room temperature.

3. PBST: PBS, 0.5% Tween-20. Store at room temperature.

4. Fetal calf serum (FCS): Store at −20 °C.

5. Blocking solution: PBS, 0.2% Tween-20, 10% FCS. 50 ml is required for each plate.

6. Citrate Buffer: 0.05 M anhydrous citric acid (Sigma), 0.1 M anhydrous sodium phosphate, adjust pH to 5.0 and filter sterilize. Store at 4 °C.

7. O-Phenylenediamine dihydrochloride (OPD, Sigma).

8. Hydrogen peroxide solution 3 wt.% in H_2O (3% H_2O_2, Sigma).

9. Stop Solution: 1 M HCl.

10. Horseradish peroxidase (HRP)-conjugated goat anti–mouse IgG (Southern Biotech).

11. 96-well flat-bottomed ELISA plates (Nunc MaxiSorp).

12. Inhalant anesthetic agents.

13. Heparinized blood collecting tubes (Fisherbrand).

14. 8 or 12 channel multichannel pipettes.

15. 50 ml conical centrifuge tubes.

16. Sonicator (Qsonica Q500-200).

17. ELISA microplate reader (Bio-Rad iMark).

2.6 ELISPOT Analysis of LCMV Specific of Plasma Cells in Bone Marrow

1. BHK21-LCMV Clone-13 lysate: Prepared in Subheading 3.1.

2. Phosphate-buffered saline (PBS): 137 mM NaCl, 2.7 mM KCl, 10 mM Na_2HPO_4, 1.8 mM KH_2PO_4, pH 7.3. Sterilize by autoclaving and store at room temperature.

3. Fetal calf serum (FCS): Store at −20 °C.

4. PBST: PBS, 0.2% Tween-20. Store at room temperature.

5. PBST-FCS: PBS, 0.2% Tween-20, 1% FCS. Store at 4 °C.

6. Roswell Park Memorial Institute (RPMI) medium-1640 basic.

7. Complete RPMI medium: RPMI-1640, 10% FCS, 2 mM L-glutamine, 100 IU/ml penicillin, 100 μg/ml streptomycin, 55 μM beta-mercaptoethanol. Store at 4 °C.

8. Red blood cell (RBC) lysis buffer: 155 mM NH_4Cl, 12 mM $KHCO_3$, and 0.1 mM EDTA. Dilute in deionized water and adjust pH to 7.3. Store at room temperature.

9. 0.1 M sodium acetate: 6.804 g sodium acetate, 500 ml deionized water. Adjust pH to 4.8–5.0.

10. 3 amino-9-ethyl-carbozole (AEC, Sigma).

11. *N,N*-dimethylformamide (DMF, Sigma).

12. AEC stock solution: 20 mg/ml AEC, DMF. Dissolve AEC in DMF, store at 4 °C in the dark for up to 1 month.

13. Biotin-anti-mouse IgG (Southern Biotech).

14. HRP-conjugated avidin-D (Vector Laboratories).

15. 75% ethanol.

16. 96-well Multiscreen plates (Millipore).

17. Dissecting instruments: Tweezers, scissors, gauze pads.

18. 1 and 5 ml syringes.

19. 70 μm strainers (BD Falcon).

20. 8 or 12 channel multichannel pipettes.

21. 50 ml conical centrifuge tubes.

22. Sonicator (Qsonica Q500-200).

23. ELISPOT plate reader (Cellular Technology Ltd., Immuno-spot® Analyzer).

3   Methods

3.1   LCMV Stock Preparation

3.1.1   LCMV Propagation

1. Recover BHK-21 cell line and subculture for 2–3 passages (*see* **Note 1**).

2. Seed 2×10^6 BHK-21 cells in T75 flask; for T175 flask, seed 4×10^6 cells. Use 15 ml of complete DMEM medium per T75 flask and 35 ml per T175 flask.

3. When the cells are approximately 70–80% confluent, decant the culture medium and infect the cells with virus of 0.03–0.1 MOI. The virus should be diluted in DMEM supplemented with 2% FCS at room temperature. For the T75 flask, use a total volume of 2 ml virus; for the T175 flask, use 4 ml (*see* **Note 2**).

4. Incubate the flasks at 37 °C for 1 h and rock 4–5 times at every 15 min interval during the infection period.

5. Add 15 or 35 ml of fresh complete DMEM medium in the T75 or T175 flask respectively and incubate at 37 °C for 48 h.

6. Harvest the medium after 48 h, and aliquot 1 ml per vial using cell frozen vials; store at −70 °C (*see* **Note 3**). Proceed to Subheading 3.1.2 for viral titration.

7. To obtain LCMV lysate, rinse the BHK-21 monolayer with PBS and pipette it out. Repeat twice (*see* **Note 4**).

8. Add 3.5 ml of PBS in a T75 flask and 7 ml in a T175 and scrape off the cells with a cell scraper. Transfer the cell suspensions into a 50 ml conical tube.

9. Sonicate cell suspensions twice at level 13 for 30 s without bubbles.

10. Dispense the LCMV lysate in 1 ml aliquots to coat the plates.

3.1.2 Quantification of LCMV Titer by Plaque Assay

1. Seed Vero E6 cells at a density of 5×10^5 cells per well of a 6-well plate in 2.5 ml of complete DMEM before 1 day prior to the plaque assay (*see* **Note 5**). Proceed with the following procedures until cells grow to 70–80% confluence.

2. Thaw the virus stock to be titrated by gentle agitation in a 37 °C water bath until ice crystals have melted and immediately place on ice.

3. Make serial tenfold dilutions of virus stock as follows: add 225 μl of DMEM 2% in each well of a 96-well round-bottomed plate; 10 wells are needed for each virus sample. Add 25 μl of virus stock in the first well and mix thoroughly by pipetting up and down to make a 10^{-1} dilution of the stock. Then, serially transfer 25 μl of diluted virus to the next well that also contains 225 μl DMEM 2%; repeat these steps to make $10^{-2 \sim -10}$ dilutions (*see* **Note 6**).

4. Completely remove the cell culture media from the 6-well plates (*see* **Note 7**).

5. Infect Vero E6 cells by adding 200 μl of virus diluents per well (*see* **Note 8**). Incubate plates at 37 °C for 60 min and rotate plates every 15 min to prevent drying.

6. During the incubation, prepare agarose overlay medium as follows: warm 2× complete medium 199 at 37 °C for at least 30 min. Microwave 1% agarose and warm at 42 °C for at least 30 min (*see* **Note 9**). Make a 0.5% agarose overlay medium by mixing together equal volumes of 2× complete medium 199 and 1% agarose.

7. Remove diluted virus away from the cells and overlay the cells with 4 ml of 0.5% agarose mix per well (*see* **Note 10**). Incubate

the 6-well plate at 37 °C for 4 days when the agarose has solidified.

8. After incubation, prepare the 0.5% agarose overlay medium as described in **step 6**. Then, add 1% neutral red in 0.5% agarose mix of a 1:50 dilution.

9. Overlay the cells with 0.5% agarose supplemented with 0.02% neutral red, 3 ml per well. Incubate plates at 37 °C for 18 h when the agarose has solidified.

10. Carefully remove the agarose layer without destroying the cell monolayer.

11. Select a well of a sufficient but not too many plaques and count the numbers of plaques (*see* **Note 11**).

12. Calculate the titer of the virus stock in plaque-forming units (pfu) per ml using the following formula: # plaques/(dilution × virus volume) = pfu/ml. For example, if 30 plaques were yielded from 0.2 ml of 10^{-8} virus dilutes, the titer of the virus is $30/(10^{-8} \times 0.2) = 1.5 \times 10^{10}$ pfu/ml.

3.2 LCMV Acute Infection of Mice

1. Thaw the virus stock by gentle agitation in a 37 °C water bath until ice crystals have melted and immediately transfer on ice; conduct all the following procedures on ice, as LCMV is heat-labile.

2. Dilute the virus stock 1:10 with the DMEM medium to obtain a virus solution of 4×10^5 pfu/ml (*see* **Note 12**).

3. Mix throughly by gently inverting the tube 6–10 times prior to infection. Infect each naive C57BL/6J mouse with 2×10^5 pfu (500 μl per mouse) virus via intraperitoneal injection.

4. House infected mice in accordance with institutional biosafety regulations.

5. All virus-contaminated equipment, including conical centrifuge tubes, 1 ml syringes, and pipettes, should be discarded in a bleach solution.

3.3 FACS Staining of GC B Cells and Plasma Cells

3.3.1 Process Lymphocytes from Spleens

1. Prepare a 60 mm × 15 mm petri dish that contains 3 ml of RBC lysis buffer and place a 70 μm cell strainer in each dish.

2. Dissect spleens from infected and naive mice. Place each spleen into the cell strainer within a dish (*see* **Note 13**).

3. Grind spleens in RBC lysis buffer using the round side of a plunger from a 1 ml syringe until only connective tissue and fat remains. Let cells pass through the cell strainer to generate single-cell suspensions.

4. Add 3 ml of RPMI 2% into the dish to dilute RBC lysis buffer.

5. Pipette the cell suspensions from the dish into a 15 ml conical tube. Rinse the dish with an additional 3 ml of RPMI 2% and transfer to the same conical tube.

6. Centrifuge the cells at $600 \times g$ for 5 min at 4 °C and discard the supernatant.

7. Resuspend the cells with 2–5 ml of RPMI 2% and count the cells.

8. Adjust the cell density to obtain approximately 2×10^7 cells per ml.

3.3.2 Process Lymphocytes from Lymph Nodes

1. Prepare 60 mm × 15 mm cell culture dishes that contain 3 ml of RPMI 2%.

2. Dissect inguinal, axillary, or brachial lymph nodes from infected and naive mice.

3. Pool lymph nodes from one mouse in one dish. Disrupt lymph nodes between frosted sides of two microscope glass slides in RPMI 2% in the dish.

4. Pipette the cell suspensions from the dish into a 15 ml conical tube. Rinse the dish with an additional 3 ml of RPMI 2% and transfer to the same conical tube.

5. Following steps refer to Subheading 3.3.1, **steps 6–8**.

3.3.3 FACS Staining

1. Dilute Fc-block 1:100 in FACS buffer for cell pre-blocking (*see* **Notes 14** and **15**).

2. Prepare a $2\times$ antibody cocktail for cell staining with antibodies listed in Subheading 2.3. Dilute antibody stocks with FACS buffer at an appropriate dilution: dilute FITC- anti-PNA at 1:50, PE-anti-FAS at 1:50, PerCP-Cy5.5-anti-CD19 at 1:100, PE-Cy7-anti-B220 at 1:100, APC-anti-CD138 at 1:50, Fixable Viability Dye eFluor® 780 at 1:500, V450-anti-IgD at 1:50. (*see* **Notes 16** and **17**).

3. Centrifuge the prepared antibody cocktails at $16,000 \times g$ for 3 min at 4 °C to remove particulates. Store prepared cocktails on ice and protect from light.

4. Load 50 μl of each sample (1×10^6 cells) to a round-bottom 96-well plate and top up with 150 μl of FACS buffer per well.

5. Centrifuge the plate for 1 min at $800 \times g$. Discard the supernatant by quickly flicking the plate. Vortex the plate for several seconds to loosen the cell pellet.

6. Wash the cells once by adding 200 μl of FACS buffer per well; centrifuge the plate for 1 min at $800 \times g$. Discard the supernatant and vortex plate.

7. Resuspend cells with 50 μl of diluted Fc-block in FACS buffer.

8. Mix well by pipetting up and down and incubate for 30 min on ice in the dark (*see* **Note 18**).

9. Add 50 μl of 2× fluorescent antibody cocktail to each well without washing.

10. Mix well by pipetting up and down and incubate for 30 min on ice in the dark.

11. Wash the cells twice by adding 200 μl of FACS buffer per well; centrifuge the plate for 1 min at 800 × *g*. Discard the supernatant and vortex the plate.

12. Resuspend the cells in 200 μl of 2% PFA to fix stained cells and transfer the cells to a 5 ml FACS tube.

13. Acquire stained samples on a flow cytometer.

3.3.4 Data Analysis and Gating Strategy

1. Load data into flow cytometry analysis software (FlowJo, Treestar).

2. Gate on lymphocytes (Fig. 1a).

3. Gate on singlets using FSC-W/FSC-H (Fig. 1b) and SSC-W/SSC-H (Fig. 1c).

4. Dump out dead cells by gating on viability dye negative cells (Fig. 1d).

5. Gate on B220$^+$CD19$^+$ B cells within single live cells (Fig. 1e).

6. Gate on PNA$^+$FAS$^+$ GC B cells within B cell population (Fig. 1f) (*see* **Note 19**).

7. Gate on IgD$^-$ cells within single live cells (Fig. 1g).

8. Gate on CD138$^+$B220$^{int/negative}$ plasma cells within IgD$^-$ cells (Fig. 1h).

3.4 FACS Staining of LCMV-Glycoprotein Specific I-A^{b}gp66-77 Tetramer T$_{FH}$ Cells

3.4.1 Tetramer Staining

1. Process the single-cell suspension from the spleens or lymph nodes of LCMV-infected and naive mice, as described in Subheadings 3.3.1 and 3.3.2.

2. Prepare tetramer staining mix by diluting I-A^{b}gp66-77 tetramer stock in complete RPMI to 2 μg/ml (*see* **Notes 20** and **21**). Vortex well and centrifuge at 16,000 × *g* for 3 min at 4 °C to pellet antibody aggregates. Store on ice and protect from light.

3. Load 100 μl of each sample (2 × 10^6 cells) to a round-bottom 96-well plate and top up with 100 μl of RPMI 2% per well.

4. Centrifuge the plate for 1 min at 800 × *g*. Discard the supernatant by quickly flicking the plate and loosen the cell pellet by briefly overtaxing the plate.

5. Resuspend each sample with 50 μl of tetramer staining mix.

6. Mix well by pipetting up and down several times and incubate for 90 min at 37 °C in the dark (*see* **Note 22**).

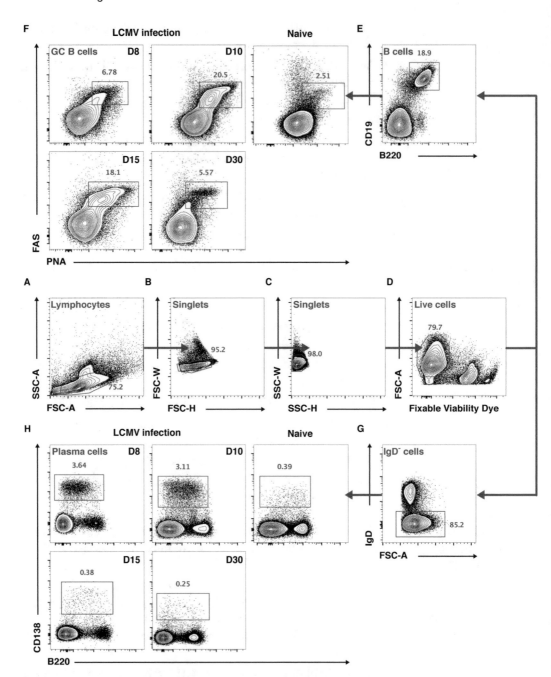

Fig. 1 Gating strategy of GC B and plasma cells. Flow cytometry staining of splenocytes from LCMV infected (8, 10, 15, and 30 days post infection) and naive mice. Gating strategy of GC B cells (**a–f**). GC B population of LCMV infected mice (**f**, *left* and *middle*) and naive mice (**f**, *right*). Gating strategy of plasma cells (**a–d**, **g–h**). Plasma cell population of LCMV infected mice (**h**, *left* and *middle*) and naive mice (**h**, *right*)

3.4.2   CXCR5 Staining

1. Prepare primary antibody (rat anti-mouse CXCR5 unconjugated) at 1:50 dilution in T_{FH} staining buffer, 50 μl per well (*see* **Note 23**). Vortex well and centrifuge at 16,000 × *g* for 3 min at 4 °C to pellet antibody aggregates. Store on ice and protect from light.

2. Wash the cells twice by adding 200 μl of FACS buffer per well; centrifuge the plate for 1 min at 800 × *g*. Discard the supernatant and briefly vortex the plate.

3. Resuspend the cells with 50 μl of primary antibody per well. Mix well by pipetting up and down and incubate for 60 min at 4 °C in the dark.

4. Wash the cells twice by adding 200 μl of FACS buffer per well; centrifuge the plate for 1 min at 800 × *g*. Discard the supernatant and briefly vortex the plate.

5. Stain with secondary antibody at 1:200 dilution (biotin-conjugated goat anti-rat IgG(H+L)) in 50 μl of T_{FH} staining buffer per well. Incubate for 30 min on ice in the dark.

6. Wash the cells twice by adding 200 μl of FACS buffer per well; centrifuge the plate for 1 min at 800 × *g*. Discard the supernatant and briefly vortex the plate.

7. Stain with streptavidin at 1:200 in 50 μl of T_{FH} staining buffer per well. At this time, include all other surface stains, such as CD4 (1:200), B220 (1:200), CD44 (1:100), ICOS (1:100), SLAM (1:100), and PD-1 (1:100) (*see* **Note 24**). Incubate for 30 min on ice in the dark.

8. Wash the cells twice by adding 200 μl of FACS buffer per well; centrifuge the plate for 1 min at 800 × *g*. Discard the supernatant and briefly overtax the plate.

9. Resuspend the cells in 400 μl of 2% PFA to fix stained cells and transfer the cells to a 5 ml FACS tube.

10. Acquire stained samples on a flow cytometer.

3.4.3   Data Analysis and Gating Strategy

1. Load data into flow cytometry analysis software (FlowJo, Treestar).

2. Gate on lymphocytes (Fig. 2a).

3. Gate on singlets using FSC-W/FSC-H (Fig. 2b) and SSC-W/SSC-H (Fig. 2c).

4. Gate on CD4$^+$ T cells and dump B cells and dead cells (CD4$^+$ B220$^-$ viability dye negative cells) (Fig. 2d) (*see* **Notes 25** and **26**).

5. Gate on CD44^{hi}gp66$^+$ cells within live CD4$^+$ T cells (Fig. 2e).

6. Gate on T_{FH} and T_{H1} cells within tetramer positive population. Determine T_{FH} population as CXCR5$^+$SLAM^{lo},

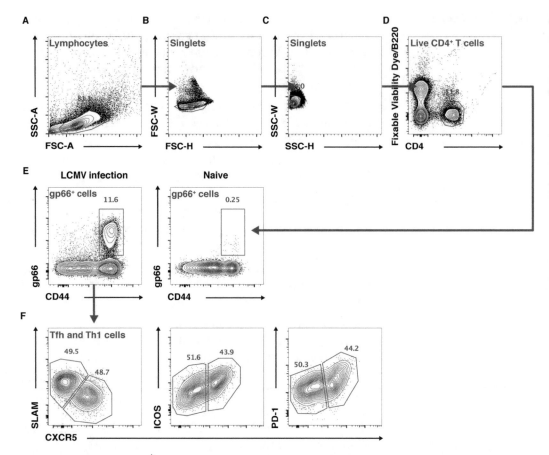

Fig. 2 Gating strategy of I-A^{b}gp66-77 tetramer specific T$_{FH}$ cells. Flow cytometry staining of splenocytes from LCMV infected (8 days post infection) and naive mice. Gating strategy of tetramer specific T$_{FH}$ cells (**a–f**). Tetramer positive CD4$^+$ T population in infected mice (**e**, *left*) and naive mice (**e**, *right*). T$_{FH}$ cell population are defined as CXCR5$^+$SLAM^{lo} or CXCR5$^+$ICOS^{hi} or CXCR5$^+$PD-1^{hi} (**f**)

CXCR5$^+$ICOS^{hi} or CXCR5$^+$PD-1^{hi}; determine T$_H$1 population as CXCR5$^-$SLAM^{hi}, CXCR5$^-$ICOS^{lo} or CXCR5$^-$PD-1^{lo} (Fig. 2f) (*see* **Note 27**).

3.5 ELISA Analysis of LCMV–IgG

3.5.1 Coat Plates with Viral Lysate and Blocking

1. Transfer the appropriate amount of BHK-LCMV lysate to a 50 ml tube and sonicate once at level 13 for 30 s (*see* **Notes 28** and **29**).

2. Dilute the lysate 1:10 in sterile PBS and coat 96-well flat-bottomed plates with 100 μl per well; wrap the plate with parafilm and incubate for 48 h at room temperature (*see* **Note 30**). Plates coated sterile may be incubated for no longer than 4 days.

3. On the day when ELISA is performed, bring all regents to room temperature prior to use and conduct all the procedures at room temperature unless otherwise specified.

4. Empty the lysate by flicking in biohazard waste and add 200 µl of PBST to each well. Allow the PBST to soak in the wells for a while and decant. Remove the residual liquid by inverting the plate and blotting it against a clean paper towel. Wash an additional 2 times with PBST (*see* **Note 31**).

5. Add 200 µl of freshly prepared blocking solution to each well and incubate for 2 h at room temperature.

3.5.2 Collect and Add Serum

1. Anesthetize the mouse and collect approximately 100 µl of blood from the orbital sinus using heparinized blood collecting tubes (*see* **Note 32**).

2. Transfer blood into a 1.5 ml EP tube; place tubes at room temperature for 30–60 min for blood coagulation.

3. Centrifuge the tubes for 15 min at $18,000 \times g$ at 4 °C.

4. Carefully aspirate serum without contaminating with red blood cells and transfer to a new EP tube.

5. Empty plate by flicking and tap out residual liquid; add 100 µl of blocking solution to each well with the exception of 150 µl for first column.

6. Add 5 µl of serum (1:30 dilution) to the first column and proceed with serial dilutions from columns 1–12. (*see* **Note 33**).

7. Make threefold serial dilutions by pipetting up and down 10 times and transferring 50 µl to the next well across the plate (*see* **Notes 34** and **35**). Discard 50 µl from the last column, thereby leaving 100 µl in each well.

8. Incubate for 90 min at room temperature.

3.5.3 Add HRP Antibody

1. Empty the plate by flicking and wash 3 times with PBST as described in Subheading 3.5.1, **step 4**.

2. Dispense 100 µl of goat anti-mouse IgG-HRP diluted 1:5000 in blocking solution to each well (*see* **Note 36**). Incubate for 90 min at room temperature. Avoid placing the plate in direct light.

3.5.4 React with Substrate

1. Empty the plate by flicking and wash 3 times with PBST as described in Subheading 3.5.1, **step 4**.

2. Prepare a substrate by dissolving 4 mg OPD (1 tablet) in 10 ml of citrate buffer. Do not touch the tablets prior to use; add 33 µl of 3% H_2O_2 to 10 ml of substrate.

3. Add 100 µl of substrate to each well.

4. After sufficient color development (approximately 8 min), add 100 µl of stop solution to each well and read the optical density (O.D.) at 490 within 30 min.

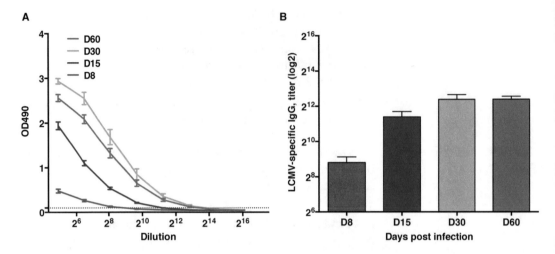

Fig. 3 ELISA analysis of LCMV–IgG. ELISA analysis of LCMV–IgG in serum from LCMV infected (8, 10, 15, and 30 days post infection) and naive mice. The results of ELISA may display as optical density (O.D.) to dilution factors (**a**) or viral titer scaled with log2 (**b**)

5. Calculate O.D. value or virus titer of each sample: plot O.D. value versus corresponding dilution factors using the mean and standard deviation of each sample (Fig. 3a). Calculate the endpoint titer at the intersection points between the OD/dilution plots and the limitation detection line determined by the O.D. value acquired from naive serum (defined as naive + 2 times the standard deviation) (Fig. 3a, b) (*see* **Note 37**).

3.6 ELISPOT Analysis of LCMV Specific of Plasma Cells in Bone Marrow

3.6.1 Coat Plates with Viral Lysate

1. Transfer the appropriate amount of BHK-LCMV lysate to a 50 ml conical tube and sonicate once at level 13 for 30 s (*see* **Note 38**).

2. Dilute the lysate 1:6.7 in sterile PBS, e.g., dilute 1.5 ml of lysate in 8.5 ml of PBS, and coat a 96-well multiscreen plate with 100 μl of diluted lysate per well; wrap the plate with parafilm and incubate for 48 h at room temperature (*see* **Note 39**).

3.6.2 Block Plates

1. On the day when ELISPOT is performed, bring all reagents to room temperature prior to use and conduct all the procedures at room temperature unless otherwise specified.

2. Empty the lysate by flicking in biohazard waste and add 200 μl of PBST to each well. Allow the PBST to soak in the wells for a while and decant. Remove the residual liquid by inverting the plate and blotting it against clean paper toweling (*see* **Note 40**). Wash an additional 2 times with PBS to remove PBST and ensure that the cells are not lysed (*see* **Note 41**).

3. Add 200 μl of freshly prepared complete RPMI medium to each well and incubate for 2 h at room temperature (*see* **Note 42**).

3.6.3 Bone Marrow Cell Preparation

1. Euthanize a mouse via cervical dislocation. Dissect femurs from mice and maintain in RPMI-1640 medium on ice until all mice have been prepared (*see* **Note 43**).

2. Remove the muscles from the bones on a gauze pad and place the clean bones in a new petri dish that contains cold RPMI medium.

3. Cut off both ends (epiphyses) of each bone with scissors.

4. Fill a 5 ml syringe fitted with a 25-G needle with 5 ml of complete RPMI medium. Insert the needle into one end of the femur and deliver 5 ml of medium to flush the marrow into a cell strainer over a 50 ml conical tube. Repeat for another femur.

5. Pass the cells through a cell strainer into the 50 ml conical tube by grinding the marrow using the round side of a plunger from a 1 ml syringe. Rinse the strainer with complete RPMI medium.

6. Centrifuge the cell suspension at $600 \times g$ for 6 min at room temperature and remove the supernatant.

7. Lyse red blood cells by adding 2–3 ml of RBC lysis buffer. Let it stand for 3 min at room temperature and quench by adding 5 ml of complete RPMI medium; centrifuge the cell suspension at $600 \times g$ for 6 min at room temperature and remove the supernatant.

8. Resuspend the cells with 1 ml of complete RPMI medium and count the cells.

9. Adjust the cell density to approximately 1×10^7 cells per ml.

3.6.4 Incubate with Cells

1. Replace old medium in a 96-well multiscreen plate with 100 μl of fresh complete RPMI medium per well to dilute the cells.

2. Add 50 μl (0.5×10^6 cells) of the single-cell suspension to the first row (*see* **Note 44**). Make serial threefold dilutions by pipetting up and down 8 times and then transferring 50 μl of cells to the next row (across or down the plate). Discard 50 μl from the last dilution so that the final volume in all the wells is 100 μl (*see* **Notes 45** and **46**).

3. Incubate the plates for approximately 8 h at 37 °C with 5% CO_2 (*see* **Note 47**).

3.6.5 Incubate with Biotin Antibody

1. After the incubation, dump off cells and wash 3 times with PBS followed by 3 times with PBST (*see* **Note 48**).

2. Add 100 μl of biotinylated anti-IgG or anti-Ig isotype of interest per well, diluted 1:500 in PBST-FCS. Incubate overnight at 4 °C (up to 2 days).

3.6.6 Incubate with HRP-Avidin Antibody

1. Dump off the antibody and wash the plate 4 times with PBST.

2. Add 100 μl of HRP-conjugated avidin-D per well 1:1000 diluted in PBST-FCS. Incubate precisely for 60 min at room temperature.

3.6.7 Develop Spots with Chromogen Substrate

1. Near the end of incubation, prepare enzyme chromogen substrate in acetate buffer: dilute AEC stock solution 1:67 into 0.1 M sodium acetate buffer, thereby providing a final concentration of 0.3 mg/ml. Filter this solution through a 0.2 μm filter and add 100 μl of 3% H_2O_2 per 10 ml of substrate prior to use.

2. Dump plate and wash 3 times with PBST, followed by an additional 3 times with PBS.

3. Add 100 μl of enzyme chromogen substrate to each well and incubate for 8 min at room temperature until red spots may be visualized and background coloring begins to become significant.

4. Stop the reaction by thoroughly washing the plate with running tap water. Remove plastic base and wash underside of plate. Blot plate dry and store in the dark away from moisture and other chemicals to prevent excess fading (color will fade over time).

5. Read plates on ELISPOT reader after fully dry to better visualize the spots.

6. Count the number of spots. True spots appear as red circular foci that are rounded by a halo-like structure (Fig. 4) (*see* **Note 49**).

D60 post LCMV infection   **Naive**

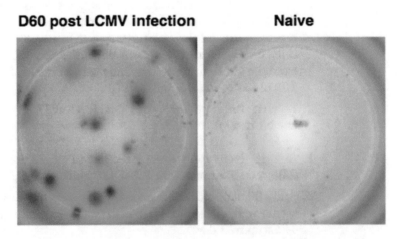

Fig. 4 ELISPOT analysis of LCMV-specific plasma cells in bone marrow. ELISPOT analysis of plasma cells in bone marrow from LCMV infected (60 days post infection) (*left*) and naive mice (*right*)

4   Notes

1. BHK-21 cells should be maintained at low passage ($<$10); cells should be adapted to propagate in culture at least 2–3 passages prior to infection.

2. It is recommended to use fresh media for cell culture and virus dilution.

3. Avoid repeat freezing-thawing because of an approximately 50% loss in virus titer in each freeze-thaw cycle [24].

4. It is necessary to use Clone-13 infected BHK cells, not Armstrong, to obtain lysate.

5. Prepare one well for each dilution of one viral sample or blank control, and an excess of several wells should be prepared in case of errors. For example, if 2 virus samples and 6 dilutions of each sample must be titrated, as well as considering a blank control and then seeding cells of at least 13 wells and an excess of 4 wells, 17 wells in total are required.

6. Always use a new pipette for each dilution.

7. Always ensure that the cells are attached to the plate as a monolayer when manipulating cells.

8. In general, an appropriate range of dilutions for testing is 10^{-4} to 10^{-10}, and it is necessary to add DMEM 2% only in well as a blank control during infection.

9. Equilibrium boiled agarose at 42 °C for at least 30 min as a high temperature will kill the Vero cells.

10. Overlay the cells as soon as possible because the agarose overlay medium turns solid very quickly, which will make it difficult to transfer medium to wells.

11. Plaques were formed with virus-infected cells that are killed by virus and fail to be stained by neutral red. Therefore, plaques are clear spots against a red or pink background when visualized by eye. It is better to confirm the real plaques with a microscope.

12. Theoretically, 1 ml of 10$\times$ stock is sufficient for the infection of 20 mice; however, a 10% excess volume of diluted virus solution should be prepared in case of an insufficient volume for the last mouse.

13. Immediately proceed to subsequent steps and avoid immersing spleens in RBC lysis buffer for a long time. Or store the isolated spleens in cold RMPI that contains 2% FBS and process splenocytes when all dissections are completed.

14. It is recommended to titrate an optimal antibody working dilution prior to experiments and adjust the titer based on values suggested by companies.

15. Calculate with 50 µl volume to be added per well of the plate. An excess of all mixes should always be prepared because pipetting errors may lead to an insufficient volume for the last sample.

16. It is available to identify the subset of GC B cells by GL-7 (T and B cell activation marker, clone GL-7) combined with FAS in flow cytometry staining.

17. If the frequency of GC B cells is too low to identify a distinct population, it may be helpful to dump immature B cells using IgD in the staining panel.

18. Avoidance of bubbles when mixing may facilitate better staining quality.

19. GC B cells are identified by high expression of CD95 (Fas) and GL-7 (T and B cell Activation Marker) [25, 26]. GC B cells bind to peanut agglutinin (PNA) and downregulate CD38 express and do not express IgD [27–30]. The early GC cell population begins to form at day 4, and the peak of the GC reaction is day 10 [2]. A distinct population of PNA$^+$FAS$^+$ GC B cells may be observed via flow cytometry at days 8–15 post LCMV infection and remains detectable at day 30.

20. It is recommended to titrate an optimal antibody working dilution prior to experiments and adjust the titer based on values suggested by companies.

21. Calculate with 50 µl volume to be added per well of the plate. An excess of all mixes should always be prepared because pipetting errors may lead to an insufficient volume for the last sample.

22. Avoidance of bubbles when mixing may facilitate better staining quality.

23. In this protocol, we recommend a three-step method based on a biotin-streptavidin system to stain CXCR5: Rat anti-mouse CXCR5 primary antibody followed by a biotinylated anti-rat IgG antibody and fluorophore-conjugated streptavidin antibody. This method enables a clearer distinction between CXCR5$^+$ and CXCR5$^-$ populations than directly using fluorophore-conjugated anti-CXCR5 antibody.

24. It is important to use a bright fluorophore-conjugated antibody to less expressed markers, such as APC, PE, or Brilliant Violet 421 conjugated to CXCR5, SLAM, or ICOS.

25. It is necessary to dump B cells when analyzing T_{FH} cells because B cells express a high level of CXCR5 that will influence the accuracy and interpretation of results.

26. CXCR5 highly expressed B cells are an ideal positive control to determine the staining quality of CXCR5.

27. It is optimal to use other markers, such as Bcl6, T-bet, TCF7, and CD44, to gate on the T_{FH} population combined with CXCR5.

28. 1 ml of lysate is sufficient for one 96-well plate.

29. It is important to repeat the sonication for 30 s prior to diluting virus lysate to coat the plate.

30. For total antibody detection, coat plates with 100 μl per well of 0.5 μg/ml goat anti-mouse IgG+IgM+IgA.

31. Correct and sufficient washing is very critical for the precision of the assay. Do not allow the wells to sit dry, and immediately proceed to the next step to reduce the differences among plates.

32. It is recommended to include a sample from a naive mouse as a negative control and a sample from a mouse infected more than 30 days as a positive control.

33. Store residual serum at −20 °C for more tests and always avoid repeated freeze-thaw cycles.

34. Multichannel pipettes are strongly recommended to use in ELISA because they will reduce errors among wells and also timesaving.

35. Pre-rinsing the pipette tips with the reagent will prevent inaccuracies caused by surface tension. Always ensure pipette tips are fitted tightly.

36. Add HRP-antibody to probe specific isotype using the anti-isotype antibody of interest.

37. Notably, it is a relative quantification for virus-specific IgG titration in this system.

38. It is important to repeat the sonication for 30 s prior to diluting virus lysate to coat the plate.

39. Plates coated sterile may be incubated for no longer than 4 days.

40. Correct and sufficient washing is very critical for the precision of the assay. Do not allow the wells to sit dry, and immediately proceed to the next step to reduce the differences among plates.

41. Thoroughly wash the plate with PBS to ensure that PBST is completely removed prior to adding cells. Remaining PBST may affect cell viability, as well as antibody binding.

42. One alternation is to prepare the plate 1 day prior to performing ELISPOT: add 200 μl of freshly prepared complete RPMI medium to each well and wrap the plate with parafilm. Store the plate at 4 °C.

43. In general, bone marrow cells from two femurs are sufficient for assay; it is not necessary to dissect tibias.

44. When performing ELISPOT, it is necessary to include a blank control by adding only complete RPMI to wells and a negative control using bone marrow cells from a naive mouse.

45. Multichannel pipettes are strongly recommended to use in ELISPOT because they will reduce errors among wells and also timesaving.

46. Pre-rinsing the pipette tips with the reagent will prevent inaccuracies caused by surface tension. Always ensure pipette tips are fitted tightly.

47. Do not disturb the plate when the cells are culturing because it may result in double spots that will not be well defined.

48. Washing with PBS ensures that remnant cells are not lysed when PBST is used.

49. Store the plates in the dark, which will prevent spot bleaching and enable spots to be analyzed over a longer period of time.

Acknowledgments

We thank Dr. Rafi Ahmed (Emory University) for generously providing us with LCMV Armstrong virus. The work was supported by National Basic Research Program of China (973 program, 2013CB531500, to L.Y.), the National Natural Science Foundation China (81471624 to L.Y.).

References

1. Plotkin SA (2008) Vaccines: correlates of vaccine-induced immunity. Clin Infect Dis 47 (3):401–409. https://doi.org/10.1086/589862

2. De Silva NS, Klein U (2015) Dynamics of B cells in germinal centres. Nat Rev Immunol 15 (3):137–148. https://doi.org/10.1038/nri3804

3. Batista FD, Harwood NE (2009) The who, how and where of antigen presentation to B cells. Nat Rev Immunol 9(1):15–27. https://doi.org/10.1038/nri2454

4. Okada T, Miller MJ, Parker I, Krummel MF, Neighbors M, Hartley SB, O'Garra A, Cahalan MD, Cyster JG (2005) Antigen-engaged B cells undergo chemotaxis toward the T zone and form motile conjugates with helper T cells. PLoS Biol 3(6):e150. https://doi.org/10.1371/journal.pbio.0030150

5. Qi H, Cannons JL, Klauschen F, Schwartzberg PL, Germain RN (2008) SAP-controlled T-B cell interactions underlie germinal centre formation. Nature 455(7214):764–769. https://doi.org/10.1038/nature07345

6. Jacob J, Kelsoe G (1992) In situ studies of the primary immune response to (4-hydroxy-3-nitrophenyl)acetyl. II. A common clonal origin for periarteriolar lymphoid sheath-associated foci and germinal centers. J Exp Med 176 (3):679–687

7. Kitano M, Moriyama S, Ando Y, Hikida M, Mori Y, Kurosaki T, Okada T (2011) Bcl6 protein expression shapes pre-germinal center B cell dynamics and follicular helper T cell heterogeneity. Immunity 34(6):961–972. https://doi.org/10.1016/j.immuni.2011.03.025

8. Kerfoot SM, Yaari G, Patel JR, Johnson KL, Gonzalez DG, Kleinstein SH, Haberman AM (2011) Germinal center B cell and T follicular helper cell development initiates in the interfollicular zone. Immunity 34(6):947–960. https://doi.org/10.1016/j.immuni.2011.03.024

9. Shih TA, Meffre E, Roederer M, Nussenzweig MC (2002) Role of BCR affinity in T cell dependent antibody responses in vivo. Nat Immunol 3(6):570–575. https://doi.org/10.1038/ni803

10. Gourley TS, Wherry EJ, Masopust D, Ahmed R (2004) Generation and maintenance of immunological memory. Semin Immunol 16(5):323–333. https://doi.org/10.1016/j.smim.2004.08.013

11. Victora GD, Nussenzweig MC (2012) Germinal centers. Annu Rev Immunol 30:429–457. https://doi.org/10.1146/annurev-immunol-020711-075032

12. Yu D, Rao S, Tsai LM, Lee SK, He Y, Sutcliffe EL, Srivastava M, Linterman M, Zheng L, Simpson N, Ellyard JI, Parish IA, Ma CS, Li QJ, Parish CR, Mackay CR, Vinuesa CG (2009) The transcriptional repressor Bcl-6 directs T follicular helper cell lineage commitment. Immunity 31(3):457–468. https://doi.org/10.1016/j.immuni.2009.07.002

13. Nurieva RI, Chung Y, Martinez GJ, Yang XO, Tanaka S, Matskevitch TD, Wang YH, Dong C (2009) Bcl6 mediates the development of T follicular helper cells. Science 325(5943):1001–1005. https://doi.org/10.1126/science.1176676

14. Johnston RJ, Poholek AC, DiToro D, Yusuf I, Eto D, Barnett B, Dent AL, Craft J, Crotty S (2009) Bcl6 and Blimp-1 are reciprocal and antagonistic regulators of T follicular helper cell differentiation. Science 325(5943):1006–1010. https://doi.org/10.1126/science.1175870

15. Schaerli P, Willimann K, Lang AB, Lipp M, Loetscher P, Moser B (2000) CXC chemokine receptor 5 expression defines follicular homing T cells with B cell helper function. J Exp Med 192(11):1553–1562

16. Kim CH, Rott LS, Clark-Lewis I, Campbell DJ, Wu L, Butcher EC (2001) Subspecialization of CXCR5+ T cells: B helper activity is focused in a germinal center-localized subset of CXCR5+ T cells. J Exp Med 193(12):1373–1381

17. Breitfeld D, Ohl L, Kremmer E, Ellwart J, Sallusto F, Lipp M, Forster R (2000) Follicular B helper T cells express CXC chemokine receptor 5, localize to B cell follicles, and support immunoglobulin production. J Exp Med 192(11):1545–1552

18. Crotty S (2011) Follicular helper CD4 T cells (TFH). Annu Rev Immunol 29:621–663. https://doi.org/10.1146/annurev-immunol-031210-101400

19. Muckenfuss RS, Armstrong C, Webster L (1934) Etiology of the 1933 epidemic of encephalitis. J Am Med Assoc 103(10). https://doi.org/10.1001/jama.1934.02750360007004

20. Zhou X, Ramachandran S, Mann M, Popkin DL (2012) Role of lymphocytic choriomeningitis virus (LCMV) in understanding viral immunology: past, present and future. Viruses 4(11):2650–2669. https://doi.org/10.3390/v4112650

21. Bocharov G, Argilaguet J, Meyerhans A (2015) Understanding experimental LCMV infection of mice: the role of mathematical models. J Immunol Res 2015:739706. https://doi.org/10.1155/2015/739706

22. Ahmed R, Salmi A, Butler LD, Chiller JM, Oldstone MB (1984) Selection of genetic variants of lymphocytic choriomeningitis virus in spleens of persistently infected mice. Role in suppression of cytotoxic T lymphocyte response and viral persistence. J Exp Med 160(2):521–540

23. Khanolkar A, Fuller MJ, Zajac AJ (2002) T cell responses to viral infections: lessons from lymphocytic choriomeningitis virus. Immunol Res 26(1–3):309–321. https://doi.org/10.1385/IR:26:1-3:309

24. Welsh RM, Seedhom MO (2008) Lymphocytic choriomeningitis virus (LCMV): propagation, quantitation, and storage. Curr Protoc Microbiol Chapter 15:Unit 15A 11. doi: https://doi.org/10.1002/9780471729259.mc15a01s8

25. Naito Y, Takematsu H, Koyama S, Miyake S, Yamamoto H, Fujinawa R, Sugai M, Okuno Y, Tsujimoto G, Yamaji T, Hashimoto Y, Itohara S, Kawasaki T, Suzuki A, Kozutsumi Y (2007) Germinal center marker GL7 probes activation-dependent repression of N-glycolylneuraminic acid, a sialic acid species involved in the negative modulation of B-cell activation. Mol Cell Biol 27(8):3008–3022. https://doi.org/10.1128/MCB.02047-06

26. Cervenak L, Magyar A, Boja R, Laszlo G (2001) Differential expression of GL7 activation antigen on bone marrow B cell subpopulations and peripheral B cells. Immunol Lett 78 (2):89–96

27. Yoshino T, Kondo E, Cao L, Takahashi K, Hayashi K, Nomura S, Akagi T (1994) Inverse expression of bcl-2 protein and Fas antigen in lymphoblasts in peripheral lymph nodes and activated peripheral blood T and B lymphocytes. Blood 83(7):1856–1861

28. Rose ML, Birbeck MS, Wallis VJ, Forrester JA, Davies AJ (1980) Peanut lectin binding properties of germinal centres of mouse lymphoid tissue. Nature 284(5754):364–366

29. Bhan AK, Nadler LM, Stashenko P, McCluskey RT, Schlossman SF (1981) Stages of B cell differentiation in human lymphoid tissue. J Exp Med 154(3):737–749

30. Oliver AM, Martin F, Kearney JF (1997) Mouse CD38 is down-regulated on germinal center B cells and mature plasma cells. J Immunol Res 158:1108–1115

Chapter 3

Expression of Exogenous Genes in Murine Primary B Cells and B Cell Lines Using Retroviral Vectors

Zhiyong Yang and Christopher D.C. Allen

Abstract

B cells, after activation, can undergo class-switch recombination and somatic hypermutation of their immunoglobulin genes, and can differentiate into memory cells and plasma cells. Expressing genes in altered versions in primary B cells and B cell lines is an important approach to understanding how B cell receptor signaling leads to B cell activation and differentiation. Recombinant retrovirus-based transduction is the most efficient method to deliver exogenous genes for expression in B cells. In this chapter, we describe streamlined protocols for using recombinant retroviral vectors to transduce both murine primary B cells and B cell lines.

Key words Retroviral vector, Spinfection, Exogenous gene expression, Primary B cell, B cell lines, Transduction

1 Introduction

After encountering cognate antigens, naive B cells will become activated and then differentiate into short-lived extrafollicular plasma cells or go on to form microanatomical structures called germinal centers, where they undergo antibody affinity maturation and differentiation into memory B cells and long-lived plasma cells. Activated B cells can also undergo class-switch recombination, through which the B cell receptor (BCR) isotype switches from IgM/IgD to IgG, IgE, or IgA. How the binding of antigens to BCRs of different isotypes translates into B cell activation and differentiation is not fully understood.

The in vivo activation of B cells in a T cell-dependent or T cell-independent type 1 manner can be mimicked by culturing primary B cells with anti-CD40 or TLR ligands, respectively, together with certain cytokines. Expressing intact or engineered forms of components of BCR signaling pathways and transcription factors such as *Prdm1* (encoding Blimp-1) in cultured primary B cells remains as an important technique to study B cell activation and

Chaohong Liu (ed.), *B Cell Receptor Signaling: Methods and Protocols*, Methods in Molecular Biology, vol. 1707, https://doi.org/10.1007/978-1-4939-7474-0_3, © Springer Science+Business Media, LLC 2018

differentiation. Transfection by electroporation and retroviral transduction are the most commonly used approaches toward this aim [1]. Using a retroviral gene delivery system, others and we have demonstrated that the mouse IgE BCR exhibits elevated activity compared to the IgG1 BCR in the absence of cognate antigen [2, 3]. Using the same approach to express chimeric B cell receptors in primary B cells, we have characterized the contribution of different domains of the IgE BCR to this antigen-independent activity [2].

Compared to primary B cells, immortalized B cell lines offer some unique advantages to study BCR signaling. For example, unlike primary B cells, some B cell lines do not need the tonic signal activity of the BCR for survival. As a result, it is possible to reconstitute BCR signaling in B cell lines by adding or removing specific components in these cells. Additionally, B cell lines can also be maintained for extended periods and can be cultured in large quantities, making them suitable for biochemical studies. Exogenous genes are usually delivered to B cell lines by chemical transfection and electroporation. Various B cell lines, such WEHI-231, BAL17, and M12 cells, can also be readily transduced by retroviruses [4]. We have studied the cell surface translocation of IgE, IgG1, and their chimeric receptors delivered to J558L cells by retroviral vectors [2].

In most cases, the recombinant retroviruses used to transduce primary B cells and B cell lines are replication incompetent. To generate such recombinant retroviruses, the gene of interest is cloned into a plasmid-based retroviral vector in place of the *gag*, *pol*, and *env* viral genes. The *gag* and *env* genes encode nucleocapsid (Gag) and envelope (Env) viral proteins respectively, while *pol* encodes protease, reverse transcriptase, and integrase. These viral proteins are necessary for retrovirus packaging and replication. To render the recombinant virus replication incompetent, the *gag*, *pol*, and *env* viral genes are either supplied in a separate plasmid vector and/or stably integrated in a packaging cell line [5]. Once a B cell is infected with recombinant retrovirus, the vector DNA with the exogenous gene will integrate into the genome of the B cell, resulting in the stable maintenance and expression of the exogenous gene in the B cell. Detailed protocols, from the construction of a retroviral vector, to the infection of murine primary B cells and B cell lines, are described below.

2 Materials

2.1 Construction of Retroviral Vector

1. Cloning vectors, such as pCR2.1 (Invitrogen).

2. Retroviral vectors, such as pQEF-T2A-Cerulean, pQCXIN (Clontech), MSCV-IRES-GFP (Addgene).

3. Packing pladmids: pCL-Eco (Addgene) or MSCV ectopic gag-pol-env plasmid (G/P/E).

4. Molecular biology reagents, such as restriction enzymes, ligase, competent cells, plasmid DNA preparation kit.

2.2 Preparation of Recombinant Retrovirus, In Vitro Culture of Primary B Cells, and Spinfection of the Cultured Primary B Cells with Retrovirus

1. Complete DMEM medium (cDMEM): DMEM high glucose medium with 10% fetal bovine serum (FBS), 10 mM HEPES, 1× penicillin/streptomycin/L-glutamine.

2. Opti-MEM reduced serum medium (Invitrogen).

3. *Trans*IT-LT1 Transfection Reagent (Mirus Bio).

4. ViralBoost (Alstem).

5. Retrovirus packaging cell line, such as Phoenix-Eco [6].

6. 5-3/4″ Pasteur pipettes.

7. DNase I (10 mg/ml, Sigma-Aldrich).

8. Anti-CD43 (clone S7) biotin (0.5 mg/ml, BD Pharmingen).

9. Anti-CD11c (clone N418) biotin (0.5 mg/ml, Biolegend).

10. ACK Lysis Buffer (Quality Biological).

11. Streptavidin MyOne T1 DynaBeads (Invitrogen).

12. Magnetic stand (Invitrogen).

13. Complete RPMI growth medium (cRPMI): RPMI 1640 medium with 10% FBS, 10 mM HEPES, 1× penicillin/strep-tomycin/L-glutamine, 50 μM β-mercaptoethanol.

14. Recombinant murine interleukin-4 (IL-4, R&D Systems or Peprotech).

15. Anti-CD40 (FGK-45, 2 mg/ml, Miltenyi Biotec).

16. Swinging bucket centrifuge that can accommodate 15/50 ml tubes and plates.

17. Tissue culture incubator.

18. Flow cytometer.

3 Methods

3.1 Construction of Retroviral-Expression Vectors

Most retroviral vectors use the inherent promoters in their long terminal repeats (LTRs) to drive the expression of exogenous genes. However, we have found that the strength of the LTR promoter can be affected dramatically by the differentiation status of B cells (*see* **Note 1**). To address the issue, we have engineered a new type of retroviral vector. The new vector, pQEF-Ceru-T2A, has a unique combination of following features (Fig. 1). First, it is derived from the pQCXIN vector, which has the U3 enhancer region deleted in the 3′LTR. In the resultant "self-inactivating" virus, the LTR is thus transcriptionally inactive, such that gene

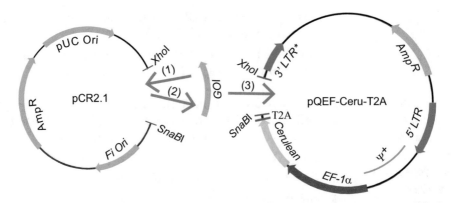

Fig. 1 Diagram of the cloning vector and the retroviral vector and the strategy for cloning exogenous genes into the retroviral vector. In *step* (*1*) the exogenous gene is cloned into the cloning vector pCR2.1 and verified by sequencing. In *step* (*2*) the exogenous gene is released from the cloning vector by restriction digestion. In *step* (*3*) the purified exogenous gene is cloned into the pQEF-Ceru-T2A retroviral vector. *GOI* gene of interest, *LTR* long terminal repeat, *3′-LTR** 3′ self-inactivating LTR, *ψ+* extended packaging signal. EF-1α: human EF-1α promoter. Amp^R, ampicillin resistance gene cassette

expression requires an inserted promoter sequence [7]. Second, for the pQEF-Ceru-T2A vector, the immediate early CMV promoter in pQCXIN, which is only weakly active in B cells, was replaced with a human EF-1α promoter [8], which allows uniform expression of exogenous genes in primary B cells irrespective of their differentiation status. Third, the pQEF-Ceru-T2A vector uses Cerulean [9], a variant of cyan fluorescent protein, as a reporter for viral infection and foreign gene expression. Finally, one or more genes of interest can be joined to the Cerulean gene by coding sequence of 2A peptides [10]. The multicistronic expression cassette will be transcribed in one transcript but the protein products of different genes will be "spliced" into individual proteins at the site of the 2A peptide during translation.

Using standard molecular biology techniques, genes of interest can be cloned into the pQEF-Ceru-T2A vector following the procedures described below and outlined in Fig. 1. A similar strategy could be used to clone genes of interest into other retroviral vectors, such as MSCV-IRES-GFP.

1. Clone desired expression cassette into a common cloning vector, such as pCR2.1. Verify the expression cassette by sequencing.

2. Clone the sequence-verified expression cassette from the cloning vector into the retroviral vector. For cloning into the pQEF-Ceru-T2A retroviral vector, the expression cassette is released from the cloning vector by digestion with restriction enzymes SnaBI and XhoI.

3. Gel purify the expression cassette, and ligate with gel-purified pQEF-Ceru-T2A that has been digested with the same restriction enzymes.

4. Transform competent *Escherichia coli* cells with the ligation sample.

5. Prepare plasmid DNA of the putative recombinant retroviral vector. Verify the identity of the clones by restriction enzyme digestion.

3.2 Preparation of Recombinant Retrovirus, In Vitro Culture of Primary B Cells, and Spinfection of the Cultured Primary B Cells with Retrovirus

We always use freshly generated retroviruses to infect mouse B cells. The whole experiment, including preparing recombinant retroviruses, setting up the B cell culture, infecting cultured B cells with the retroviruses, and analyzing the transduced B cells, takes about 6 days. Some of these parts of the experiment are overlapping. For better planning and executing the experiment, we find it is helpful to summarize workflow of the experiment in a timeline flowchart (Fig. 2). The detailed protocol is described below, using a 12-well plate to culture the packaging Phoenix-Eco cells, and a U-bottom 96-well plate to culture B cells as examples (*see* **Note 2**).

3.2.1 Preparation of Recombinant Retroviruses

1. (Day 0) Seed 0.35 million Phoenix-Eco cells in 1 ml complete DMEM medium per well, so that the cell density will be at about 50%–70% confluency on the next day.

2. (Day 1) Transfect the Phoenix-Eco cells. Add 0.7 µg retroviral plasmid DNA, 0.3 µg G/P/E plasmid DNA to 100 µl Opti-MEM reduced serum medium in a 1.5 ml Eppendorf tube, then add 3 µl *Trans*IT-LT1 Transfection Reagent (*see* **Note 3**). Mix and incubate at room temperature for 15–30 min. Then add the *Trans*IT-LT1:DNA complex mixture dropwise to the cell culture.

3. (Day 2) Replace the medium of the transfected Phoenix-Eco cells with fresh medium in the morning, and again replace the medium with new medium with 1× ViralBoost in the evening (*see* **Note 4**).

Day 0	Day 1	Day 2	Day 3	Day 4-6
Seed Phoenix cells	Transfect Phoenix cells	Change media.	Prepare retroviral suspension	
		Set up splenocyte or B cell culture	Spinfect the cultured splenocytes or B cells	Analysis by flow cytometry etc.

Fig. 2 Workflow for preparing recombinant retroviruses and transducing splenocytes or B cells by spinfection. Tasks are grouped into distinct days. Tasks involving handling of Phoenix-Eco cells are listed above the line, whereas those involving handling of splenocytes/B cells are listed beneath the line

1. (Day 2) Euthanize a mouse of the desired genotype, harvest the spleen into a 15 ml conical tube with 5 ml cDMEM medium on ice. Mash the spleen on a 40 μm cell strainer in a 10 cm Petri dish using the bulb of a 3 cc syringe. Pipette the cell suspension three times through a 5-3/4″ Pasteur pipette to break tissue particles. Then pass the cell suspension through the cell strainer again to remove unbroken cell aggregates. Centrifuge the cell suspension at $400 \times g$ at 4 °C for 5 min.

2. Remove the supernatant and add 10 μl DNase I (10 mg/ml) to the cell pellet, then resuspend the cell pellet in 1 ml cDMEM by pipetting slowly and carefully using a P1000 pipette.

3. Count the cells.

4. Add biotin conjugated anti-CD43 and anti-CD11c antibodies to the splenocyte sample. Use 10 μl of each antibody per 1×10^8 cells. Tap the tube to mix, and incubate 25–60 min on ice with gentle mixing every 10–15 min.

5. Lyse red blood cells by adding 9 ml ACK Lysis Buffer at room temperature to the tube. Invert tubes to mix. Spin at $400 \times g$ at 4 °C for 7 min.

6. Meanwhile, prepare Streptavidin MyOne T1 DynaBeads:

   (a) Resuspend beads by pipetting up and down and gentle vortexing.

   (b) Transfer 250 μl beads per 1×10^8 cells into a 1.5 ml Eppendorf tube.

   (c) Add 1 ml cDMEM media per tube to wash.

   (d) Put the tube on a magnet stand, wait 2 min. The beads will bind to the sides of the tube. Aspirate media.

   (e) Remove the tube from the magnet, resuspend beads in the original volume of cDMEM.

7. Take the 15 ml tube with cells out of centrifuge, aspirate the supernatant, and add 10 μl DNase I to the cell pellet. Slowly, carefully resuspend cells in 1 ml media using a P1000 pipette.

8. Transfer the cells to the Eppendorf tube with the washed beads. Invert the tube to mix, incubate on a tube rotator at 4 °C for 10 min.

9. Put the Eppendorf tube on a magnet stand at room temp for 2 min. The beads will bind to the sides of the tube.

10. Carefully remove negative fraction (without beads) using a short Pasteur pipette and transfer to a new 1.5 ml Eppendorf tube.

11. Centrifuge at $400 \times g$ at 4 °C for 2 min. Remove the supernatant, resuspend the cell pellet in 0.2 ml cRPMI.

12. The purity of the cell sample can be assessed by flow cytometry after staining with fluorescently conjugated anti-CD19, anti-CD11c, and anti-CD3e antibodies. The purity of B cells using this protocol is usually above 95%.

*3.2.3 Setting Up Splenocyte/B Cell Culture (See **Notes 6** and **7**)*

1. (Day 2) Add 100 µl cRPMI with 2× final concentration of IL-4 and anti-CD40 to each well of a U-bottom 96-well plate.

2. Then add 100 µl cRPMI with 100,000 splenocytes or purified B cells to the same wells of **step 1** (*see* **Note 8**).

3. Place the culture in a humidified tissue culture incubator at 37 °C with 5% CO_2.

*3.2.4 Spinfection (See **Note 9**)*

1. (Day 3) Transfer the medium of the transfected Phoenix-Eco cells into new tubes, centrifuge $1000 \times g$ at room temperature for 2 min. Transfer the supernatant containing retroviral particles into new tubes, add 1/100 volume of 1 M HEPES, and polybrene to a final concentration of 5 µg/ml (*see* **Note 10**).

2. Centrifuge the plate with cultured B cells at $750 \times g$ at room temperature for 5 min. Carefully transfer the growth media to wells in a new plate. Save the plate in the same tissue culture incubator.

3. Immediately add 195 µl/well viral supernatant to the wells with B cell pellet. Centrifuge at $1100 \times g$ at room temperature for 90 min.

4. Aspirate the supernatant after spinfection, and immediately add the growth media saved in **step 2** back to the cell pellet.

5. Return the plate(s) back to 37 °C incubator for further culture until the time to be analyzed.

*3.2.5 Spinfect B Cell Lines with Recombinant Retroviruses (See **Note 11**)*

The protocol for infecting B cell lines with retroviruses is almost identical to the one for infecting primary B cells. The major difference is that 10,000 cells in logarithmic growth phase can be directly added to each well of a U-bottom 96-well plate for spinfection.

4 Notes

1. Retroviral plasmid vectors that use the intrinsic promoters residing in the LTRs of Moloney murine leukemia virus (MoMLV) or murine stem cell virus (MSCV) have been widely used for generating recombinant retroviruses for B cell infection. However, we have found that the expression of exogenous genes under the LTR promoter varies widely, with nearly an order of magnitude higher expression in plasma cells (PCs) compared with non-PCs (Fig. 3). In some cases, the low expression of exogenous genes in non-PCs could potentially make it difficult to clearly separate the transduced cells from

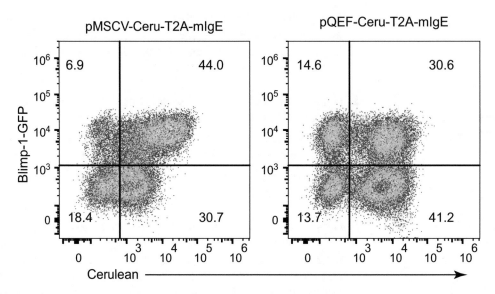

Fig. 3 Comparison of expression of an exogenous gene driven by different promoters. The bicistronic expression cassette Cerulean-T2A-mIgE was cloned into a MSCV-based vector under the endogenous promoter in the LTR (pMSCV-Ceru-T2A-mIgE) or in pQEF-Ceru-T2A retroviral vector under the EF-1a promoter (pQEF-Ceru-T2A-mIgE). Splenocytes from Blimp1-gfp knock-in reporter mouse were spinfected with the resultant recombinant retroviruses after 2 days of in vitro culture with IL-4 and anti-CD40 and then analyzed by flow cytometry after 2 more days of in vitro culture. Cerulean was used as a reporter for viral transduction and exogenous gene expression. GFP was used as a reporter for Blimp-1 expression and plasma cell differentiation

non-transduced cells. To address these issues, we replaced the CMV promoter in the pQCXIN vector with EF-1α promoter. Additionally, we cloned a Cerulean-T2A expression cassette immediately downstream of the EF-1α promoter so that the fluorescent protein Cerulean can be used as a reporter for retroviral infection and exogenous gene expression. The resultant new retroviral vector, pQEF-Ceru-T2A (Fig. 1), allows relatively uniform expression of exogenous genes in both PCs and non-PCs, and all the transduced cells can be unequivocally separated from the untransduced cells (Fig. 3).

2. Several categories of reagents, such as calcium phosphate, and liposome-based reagents, have been successfully used for transfecting the Phoenix-Eco packing cells. Among these reagents, we have found that *Trans*IT-LT1 Transfection Reagent has several advantages. First, it has low toxicity to cells and does not need to be removed several hours after transfection. Second, it works well even in a small volume of the transfection mixture, and is thus suitable for preparing tens of different transfection mixtures in parallel.

3. The protein products of retroviral genes *gag*, *pol*, and *env* are necessary for the replication and packaging of recombinant retroviruses. These genes have been stably integrated into the

Phoenix-Eco packaging cells. However, the Phoenix-Eco cells tend to gradually lose these genes with extended passaging. So it is advisable to use Phoenix-Eco cells with fewest passages possible. Additionally, we have found that co-transfection of a plasmid containing the MSCV ecotropic gag-pol-env DNA together with the retroviral vector DNA improves the yield of recombinant viruses. We co-transfect the MSCV ecotropic gag-pol-env plasmid DNA and the retroviral DNA at a mass ratio of 3:7.

4. Addition of ViralBoost is optional. We have observed an approximately two-fold increase in the expression of retrovirally-delivered exogenous genes with $1 \times$ ViralBoost.

5. We have found that unpurified splenocytes and white blood cells can be used in place of purified B cells for most applications [11]. In the presence of anti-CD40 and IL-4, B cells are the primary cell type to undergo activation and proliferation whereas most other cells die out. Purified B cells should be used in applications in which the contribution of non-B cells needs to be excluded or for analyses of total cell lysates, such as measurements of mRNA or protein abundance.

6. We have observed that the source of FBS makes a substantial qualitative difference in the success of in vitro B cell cultures. Thus, it is advisable to screen different lots of FBS, possibly from different commercial sources, to identify the FBS that allows the most robust B cell culture. Then the FBS from the same lot can typically be acquired in affordable quantities and stored in a $-80\ ^{\circ}C$ freezer for long-term use.

7. We have found that the potency of anti-CD40 antibody from different sources varies dramatically (even with supposedly the same clone). In addition, the concentration of anti-CD40 antibody can have a significant impact on B cell differentiation [2]. We usually use 125 ng/ml of anti-CD40 (clone FGK-45, Miltenyi Biotec) for culturing B cells.

8. We suggest setting up B cell culture plates using "two component" system, one component with a $2 \times$ concentration of activation reagents and the second component with cells in media. This approach is convenient, especially under conditions where multiple treatments or multiple sources of cells are desired.

9. A previous protocol concentrated retrovirus by centrifugation of viral-containing supernatant and then used the concentrated retrovirus to transduce B cells [12]. The transduction efficiency for primary B cells using that protocol was reported to be 20%–50%. With the protocol described in this chapter, and without the steps to concentrate retroviruses, we can routinely achieve 30%–50% transduction efficiency for primary B cells. Under ideal conditions, the transduction efficiency for primary B cells can be as high as 90%.

10. Before being added to B cell pellets to be infected, individual retroviral supernatants can be placed in the wells of a 96-deep well plate. This allows transfer up to 12 samples simultaneously with a multichannel pipette.

11. B cell lines are invaluable to study B cell biology. These cells are usually either transfected by electroporation [13] or by retroviral vectors [14]. Compared to transient transfection by electroporation, transduction by retroviral infection has at least two advantages. First, the retroviral method does not require an electroporator. Second, the infected cells are stably "transfected," by definition, due to retroviral integration. However, we observed that the amenability of B cell lines to retroviral infection varies dramatically, from close to 0% of A20 cells to nearly 100% of J558L cells (Fig. 4). Therefore, it is advisable to test the infectivity of B cell lines by retroviruses if the retroviral transduction approach is desired for future experiments.

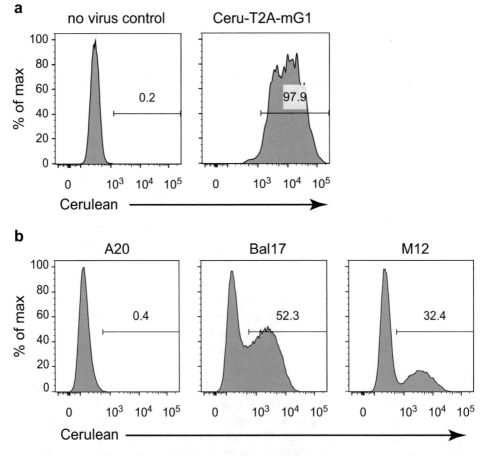

Fig. 4 Examples of expression of exogenous genes delivered by retroviral vectors in B cell lines. Histogram of the Cerulean reporter expressed in J558L cells infected with or without viruses (**a**) and A20, Bal17 and M12 cells infected with viruses (**b**). The retroviruses used in the experiments were generated from pQEF-Ceru-T2A-mIgG1 vector that expresses membrane IgG1

Acknowledgments

These protocols were developed for research projects supported by the UCSF Sandler Asthma Basic Research Center, the UCSF Cardiovascular Research Institute, the Weston Havens Foundation, and grant DP2HL117752 from the National Institutes of Health. C.D.C.A. is a Pew Scholar in the Biomedical Sciences, supported by The Pew Charitable Trusts. The content is solely the responsibility of the authors and does not necessarily represent the official views of the National Institutes of Health or of the Pew Charitable Trusts.

References

1. Moghimi B, Zolotukhin I, Sack BK, Herzog RW, Cao O (2011) High efficiency ex vivo gene transfer to primary murine B cells using plasmid or viral vectors. J Genet Syndr Gene Ther 2(103). https://doi.org/10.4172/2157-7412.1000103

2. Yang Z, Robinson MJ, Chen X, Smith GA, Taunton J, Liu W, Allen CD (2016) Regulation of B cell fate by chronic activity of the IgE B cell receptor. eLife 5:e21238. https://doi.org/10.7554/eLife.21238

3. Haniuda K, Fukao S, Kodama T, Hasegawa H, Kitamura D (2016) Autonomous membrane IgE signaling prevents IgE-memory formation. Nat Immunol 17(9):1109–1117. https://doi.org/10.1038/ni.3508

4. Krebs DL, Yang Y, Dang M, Haussmann J, Gold MR (1999) Rapid and efficient retrovirus-mediated gene transfer into B cell lines. Methods Cell Sci 21(1):57–68

5. Chang T, Yee JK (2012) General principles of retrovirus vector design. Methods Enzymol 507:1–14. https://doi.org/10.1016/B978-0-12-386509-0.00001-6

6. Pear WS, Scott ML, Nolan GP (1997) Generation of high-titer, helper-free retroviruses by transient transfection. Methods Mol Med 7:41–57. https://doi.org/10.1385/0-89603-484-4:41

7. Yu SF, Vonruden T, Kantoff PW, Garber C, Seiberg M, Ruther U, Anderson WF, Wagner EF, Gilboa E (1986) Self-inactivating retroviral vectors designed for transfer of whole genes into mammalian-cells. Proc Natl Acad Sci U S A 83(10):3194–3198. https://doi.org/10.1073/pnas.83.10.3194

8. Wakabayashi-Ito N, Nagata S (1994) Characterization of the regulatory elements in the promoter of the human elongation factor-1 alpha gene. J Biol Chem 269(47):29831–29837

9. Rizzo MA, Springer GH, Granada B, Piston DW (2004) An improved cyan fluorescent protein variant useful for FRET. Nat Biotechnol 22(4):445–449. https://doi.org/10.1038/nbt945

10. Szymczak-Workman AL, Vignali KM, Vignali DA (2012) Design and construction of 2A peptide-linked multicistronic vectors. Cold Spring Harb Protoc 2012(2):199–204. https://doi.org/10.1101/pdb.ip067876

11. Yang Z, Sullivan BM, Allen CD (2012) Fluorescent in vivo detection reveals that IgE(+) B cells are restrained by an intrinsic cell fate predisposition. Immunity 36(5):857–872. https://doi.org/10.1016/j.immuni.2012.02.009

12. Lin KI, Calame K (2004) Introduction of genes into primary murine splenic B cells using retrovirus vectors. Methods Mol Biol 271:139–148. https://doi.org/10.1385/1-59259-796-3:139

13. Liu W, Chen E, Zhao XW, Wan ZP, Gao YR, Davey A, Huang E, Zhang L, Crocetti J, Sandoval G, Joyce MG, Miceli C, Lukszo J, Aravind L, Swat W, Brzostowski J, Pierce SK (2012) The scaffolding protein synapse-associated protein 97 is required for enhanced signaling through isotype-switched IgG memory B cell receptors. Sci Signal 5(235):ra54. https://doi.org/10.1126/scisignal.2002820

14. Lee AH, Iwakoshi NN, Anderson KC, Glimcher LH (2003) Proteasome inhibitors disrupt the unfolded protein response in myeloma cells. Proc Natl Acad Sci U S A 100(17):9946–9951. https://doi.org/10.1073/pnas.1334037100

Chapter 4

Biophysical Techniques to Study B Cell Activation: Single-Molecule Imaging and Force Measurements

Ivan Rey, David A. Garcia, Brittany A. Wheatley, Wenxia Song, and Arpita Upadhyaya

Abstract

Cells of the adaptive immune system recognize pathogenic peptides through specialized receptors on their membranes. The engagement of these receptors with antigen leads to cell activation, which induces profound changes in the cell including cytoskeleton remodeling and membrane deformation. During this process, receptors and signaling molecules undergo spatiotemporal reorganization to form signaling microclusters and the immunological synapse. The cytoskeletal and membrane dynamics also leads to exertion of forces on the cell-substrate interface. In this chapter we describe two techniques—one for single-molecule imaging of B cell receptors to measure their diffusive properties as cells get activated on supported lipid bilayers; and the second for visualizing and quantifying cellular forces using elastic surfaces to stimulate T and B cells.

Key words B cell, B cell receptor, Signaling, Single-molecule imaging, Traction force microscopy, Cytoskeletal forces, Substrate stiffness

1 Introduction

The activation of lymphocytes is an essential step in the adaptive immune response [1]. Lymphocyte activation involves the binding of specialized receptors (TCR in T cells and BCR in B cells) with antigen on the surface of antigen-presenting cells. This leads to changes in cell morphology and the movement and assembly of receptors, scaffold proteins and enzymes into signaling microclusters, which are essential for immune cell activation [2]. During this process, the cells of the immune system interact with structures that possess a diverse range of physical properties. In response to stimulation by an antigen-presenting surface, B cells undergo rapid initiation of signaling for 3–5 min, concurrent with cell spreading and formation and growth of microclusters [3]. This is followed by a later stage of signaling down regulation (7–10 min), which is accompanied by cell contraction, microcluster coalescence into a

Chaohong Liu (ed.), *B Cell Receptor Signaling: Methods and Protocols*, Methods in Molecular Biology, vol. 1707, https://doi.org/10.1007/978-1-4939-7474-0_4, © Springer Science+Business Media, LLC 2018

central cluster, and eventually internalization of antigen by endocytosis [4]. Activation leads to cytoskeletal dynamics, which are critical in the exertion of mechanical forces that support signaling activation, receptor movement, and the assembly of signaling microclusters [5].

The formation of signaling microclusters depends critically on the movement of individual B cell receptors, their interactions with other receptors or signaling molecules, actin dynamics and actin regulatory proteins, as well as the nature of the local physical membrane environment [6–12]. In order to decipher the fundamental mechanisms underlying microcluster assembly, it is important to determine single BCR movements and how they respond to cytoskeletal forces and dynamics. These receptor movements at the molecular scale are accompanied by micron scale reorganization of the actin cytoskeleton, membrane deformations which exert forces at the cell interface. These forces in turn may facilitate receptor movement or triggering (by deformations or conformation changes). Recent work in the field has examined from a biophysical perspective how T cells and B cells respond to physical cues such as stiffness and ligand mobility [5, 6, 13]. Here, we present two techniques to examine the dynamics of B cells during the immune response at two scales—using single-molecule imaging to quantify receptor movement at the molecular scale, and using traction force microscopy to measure forces exerted by cells at the cellular scale.

Fluorescence microscopy has allowed cell biologists to observe, study, and gain a better understanding of the role of specific proteins in different cellular processes. Little was known of the dynamics of cellular processes at the nano-scale until the advent of localization microscopy because standard optical microscopy cannot resolve two features within 250 nm and the cell is a crowded environment [14–16]. By lowering the density of molecules imaged at a time by, for example, stochastical photoactivation [15, 17, 18], it is possible to improve the localization precision of single emitters down to a few tens of nanometers by adjusting a Gaussian fit to their intensity profile [19]. The development of these techniques allowed the mapping of cellular structures at the molecular level. Since the advent of super-resolution imaging by localization, several different techniques have been developed. Many of these techniques are aimed to discern structures and are best suited for fixed samples; however, some techniques have been adapted to allow studying the dynamics at the molecular level on living cells and organisms. Single-particle tracking (SPT) allows the study of the nano-scale properties of the cell through the measurement of parameters like the diffusion constant, among the several contributions to highlight, SPT studies of the cell membrane changed our perspective on the cell membrane from the fluid mosaic model to a more heterogeneous and compartmentalized structure [20].

The dynamics of B cell receptors (BCR) has been studied before at the single-molecule level using synthetic systems to mimic the surface of the antigen-presenting cell (APC). Both glass [6, 10] and supported bilayer substrates [4] have been used to activate the cells and track the movement of individual molecules. The labeling of the BCR is typically performed by incubating the cells in a solution containing the fluorescent antibody at low concentrations and at low temperatures to avoid antibody internalization. While this approach allows visualization of single molecules, it usually restricts the extent of time over which BCR dynamics can be tracked due to photobleaching or loss of resolution due to cell contraction and receptor coalescence. In this chapter, we describe a technique that allows imaging of BCR dynamics at single-molecule resolution for extended periods of time (>10 min) on B cells activated on a supported lipid bilayer, in order to study different stages of signaling from activation to downregulation.

Concurrent with receptor movement, the cell membrane spreads outward driven by actin polymerization and exerts forces on the cell-surface receptors as they engage ligand. These forces may facilitate signaling activation by inducing conformational changes in the receptor-ligand complexes and regulating microcluster assembly and movement. Recent work has shown that T cell activation is dependent on physical forces and B cell antigen discrimination and internalization are modulated by forces [21]. Further, both T and B cell signaling has been shown to be dependent on the stiffness of the antigen-presenting surface [13]. In the second part of this chapter, we present a technique–traction force microscopy, to measure the forces exerted by lymphocytes [22]. Cells are activated on soft elastic surfaces with beads as fiduciary markers to monitor the displacement of the gel in response to cellular forces [23]. The exerted traction stresses are calculated from the displacement field of the beads and the known elastic modulus of the gel [23–25]. A major advantage of this technique is that the gel stiffness can be modulated to study the role of substrate stiffness on cell behavior and signaling. We describe the procedure to make elastic gels of tunable stiffness by varying the ratio of the polymer acrylamide to its crosslinker bis, to coat the gel with antigen of interest, measure gel stiffness, and quantify cellular forces.

2   Materials: Single-Molecule Imaging

For the preparation of all solutions use ultrapure water and keep the prepared stocks at 4 °C.

2.1 Liposome and Antigen Preparation

1. DiH$_2$O.
2. DiH$_2$O 70% Ethanol.
3. NanoStrip, a stabilized formulation of sulfuric acid and hydrogen peroxide.
4. PBS 1× filtered (0.50 μm pore size).
5. 1,2-dioleoyl-sn-glycero-3-phosphocholine (DOPC) 20 mg/mL (Avanti Polar Lipids).
6. 1,2-dioleoyl-sn-glycero-3-phosphoethanolamine (DOPE)-cap-biotin 25 mg/mL (Avanti Polar Lipids).
7. Goat anti-mouse IgM Fab2' fragments (Jackson Immunoresearch).
8. Maleimide-PEG2-Biotin (Thermo Scientific).
9. Alexa Fluor 546 (AF546) labeling kit (Invitrogen).

2.2 Spleen Harvest of Primary B Cells and Imaging

1. DMEM (Lonza).
2. DMEM + 5 FBS (Lonza).
3. Histopaque (Sigma).
4. CD90 (Thy1) anti-mouse (Invitrogen).
5. 1× HBSS (Gibco).
6. Trypan Blue (Gibco).
7. No 1.5 24 × 50 mm coverslip (Fisherbrand).
8. 8-well Plastic Chamber (Nunc Labtek).
9. Leibovitz L-15 medium + 2% FBS (Gibco).

2.3 Equipment

1. Inverted microscope Nikon TE2000 PFS.
2. iXon X3 EMCCD camera (Andor).
3. High numerical aperture lens N.A 1.49 100× or 60×.
4. 1.5× magnification set.
5. Laser line of 546 nm or similar wavelength (for red fluorophore excitation) 20 mW or higher (Cobolt).
6. Acousto-optic tunable filter (AOTF) (Andor).
7. Filter wheel and filter set for red-range fluorophore emission (Sutter).
8. Computer with software for microscope operation that allows stream-mode imaging.

---

3 Methods: Single-Molecule Imaging

3.1 Liposome Preparation

Liposome preparation is performed by following a well-established protocol previously described [31]. A summary of the main steps is presented below.

1. Prepare a 100:1 molar ratio of 1,2-dioleoyl-sn-glycero-3-phosphocholine and 1,2- dioleoyl-sn-glycero-3-phosphoethanolamine-cap-biotin (Avanti Polar Lipids) in filtered 1× PBS (5 mM total concentration).

2. Sonicate the lipid mixture until it becomes clear (the lipid mixture will look cloudy at the beginning).

3. Centrifuge and filter to get rid of any aggregates.

3.2 Biotinylation and Labeling of Antigen

For the production of monobiotinylated goat anti-mouse IgM antibodies a protocol previously described [26] is followed. Here we summarize the steps:

1. The disulfide bonds of Fab2′ fragments (Jackson Immunoresearch) are reduced by 2-Mercaptoethylamine (Thermo Scientific).

2. The resulting sulphydryls are then biotinylated with Maleimide-PEG2-Biotin (Thermo Scientific).

3. To determine if the ratio of biotin to Fab′EZ Biotin is 1:1, the Pierce Streptavidin Plus UltraLink quantitation kit (Pierce Protein) is used.

4. The mono-biotinylated fragments of antibody (mbFab) can then be fluorescently labeled using the AF546 labeling kit (Invitrogen).

3.3 Materials for Bilayer Preparation

1. Leave glass slides overnight in Nanostrip solution and then blow with filtered air to dry. Glue to a plastic 8 wells chamber (Nunc Labtek) using a silicon elastomer (Sylgard).

2. Prepare and filter 500 mL of PBS 1×.

3. Prepare a 10 μM liposome solution by mixing 20 μL of DOPC/DOPE-cap-biotin (5 mM) liposome solution with 980 μL of PBS 1×.

4. Prepare 2 μg/mL solution of streptavidin by mixing 2 μL of streptavidin at 1 mg/mL with 998 μL of PBS 1×.

5. Prepare 18 μg/mL solution of unlabeled mono-biotinylated fragment of antibody (mbFab).

6. Prepare 0.05 mg/mL solution of AF546 labeled mbFab.

3.4 Preparation of Lipid Bilayer

Keep all solutions and reagents on ice while not in use to extend their shelf life.

1. Add 250 μL of prepared liposome solution to each well, making sure to cover the entire bottom surface. Incubate for 10 min, covered, at room temperature.

2. Carefully wash each well with 20 mL of 1× PBS (filtered). Do this by inserting the tip of the aspirating pipette and the tip of

the pipette with PBS diagonally across from one another pointing toward the upper corners of each well.

3. After washing, add PBS to each well and aspirate so that the well is full and the liquid is continuous with the top of the well. The total volume of each well is 700 μL. Remove 450 μL of PBS from each well by carefully placing micropipette tip against a lower corner (without touching the bottom) to leave a volume of 250 μL.

4. Add 250 μL of prepared streptavidin solution to each well and incubate, covered, at room temperature for 10 min. Repeat **steps 2** and **3**.

5. Add 250 μL of prepared mbFab solution to each well and incubate, covered, at room temperature for 10 min. Repeat **steps 2** and **3**.

6. Wells should be left with 250 μL volume of PBS, covered and stored at 4 °C until use.

3.5 Spleen Harvest for Primary B Cells

1. Warm reagents in 37 °C water bath before use: DMEM, Ficoll (Histopaque), 1× HBSS, 5% complete media (CM). Prepare mouse and dissection pan, then remove spleen and transfer to warm DMEM. Macerate the spleen using frosted-edge slides into petri dish with DMEM, working in small fragments until no large pieces remain.

2. Transfer macerated splenic cells to conical tube and centrifuge at $660 \times g$ to pellet all cells in suspension. Remove the supernatant, resuspend well in DMEM and underlay with warm Ficoll to create density gradient layer. Centrifuge at $1179 \times g$ for 20 min to separate red blood cells from splenic white blood cells in suspension. Carefully extract the top and middle layers above Ficoll and transfer to a new conical tube. Take a small volume and mix with Trypan blue for cell count.

3. Spin the conical tube at $660 \times g$ to pellet B cells and white blood cells. In the meantime, count cells and perform the following calculations for the next resuspension. For volume (in mL) required of 1× HBSS, divide the total number of cells by 4×10^7. Divide mL of HBSS by 10 to get the required volume of complement (in μL). For volume of Thy-1 antibody (in μL), multiply mL of HBSS by 2. After centrifugation, remove the supernatant and resuspend in calculated volumes of HBSS, complement and Thy-1 antibody. Mix well. Place the conical tube in 37 °C water bath for 30 min, mixing and inverting tube every 10 min. Heating will activate complement, which will recognize and lyse Thy-1 bound T cells.

4. Fill conical tube to 15 mL with HBSS, spin at $660 \times g$ to pellet remaining B cells and monocytes. Remove the supernatant and resuspend in 10 mL complete media. Transfer to cell culture

flask. Rinse conical flask twice with 10 mL complete media, again transfer to culture flask so that total volume is 30 mL. Place the flask in 37 °C + CO$_2$ incubator for 1 h to pan, i.e., the remaining T cells and monocytes will adhere to abiotic flask surfaces while B cells will not.

5. Transfer the cells to 50 mL conical tube and spin at 660 × g to pellet primary B cells. Remove the supernatant, resuspend cells in DMEM, and mix a small sample with Trypan Blue for cell count. Centrifuge again at 660 × g and resuspend pelleted cells in a small volume of DMEM for an ideal concentration of 12 × 10^6 cells/mL.

6. Place the cells on ice (at 4 °C) soon after obtaining desired concentration. Keep the cells on ice to extend their usable time (4–6 h).

3.6 Imaging of B Cell Receptors

For the study of B cell receptor dynamics we use an inverted microscope (Nikon TE2000 PFS) equipped with a 1.49 NA 100× lens and an electron multiplied CCD (EMCCD) camera (Ixon from Andor). The high numerical aperture objective allows TIRF illumination, which due to the short penetration depth of the evanescent field (~100 nm) improves signal-to-noise ratio for observing movements of molecules at the cell membrane. In order to image single molecules on the cell membrane for extended periods of time, we add the fluorescent antibody in solution as shown in Fig. 1. The strategy of labeling while imaging has been previously implemented by the technique uPAINT [14, 27], where the fluorescent markers in solution enabled obtaining a larger number of particle tracks and improved statistics.

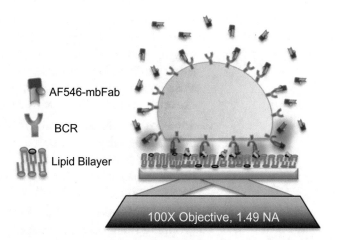

Fig. 1 Experimental setup for B cell activation on a supported lipid bilayer. The fluorescent markers for BCR in solution allow imaging the dynamics of the molecules at many different time points by replenishing the cell contact area

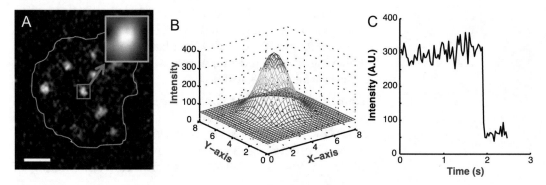

Fig. 2 Single molecule detection. (**a**) Single frame showing several fluorescently labeled mbFab on a cell labeling a few B cell receptors. Scale bar 1 μm. (**b**) Gaussian fit for the magnified molecule enabling sub-diffractive localization. (**c**) Plot of intensity of the same detected molecule over time displaying the typical single step photo-bleaching of single fluorophores as a sudden drop in intensity

1. Replace PBS in wells by L-15 + 2% FBS. Don't allow surface to dry.

2. Add 0.75 μL of AF546-mbFab (0.05 mg/mL) solution in each well to obtain a final concentration of 0.15 nanograms per milliliter (ng/mL) of fluorescent antibody. This concentration will allow for detection of single-molecule events as shown in Fig. 2

3. Use a photometer to determine the microscope-computer settings that provide a laser intensity of 5 mW at the back focal plane of the objective for a 546 nm laser line.

4. Add immersion oil on the objective and place the 8-well chamber on microscope stage. Use interference reflection microscopy (IRM) to focus on glass surface if available. Turn on the perfect focus system (or a similar focus maintaining system).

5. Set up the imaging parameters for the fluorophore of interest. These include the appropriate filters, camera exposure, and imaging mode. We use the software suite, Andor iQ, for acquisition and imaging is done in stream mode. This allows imaging at 33 frames per second using the whole camera sensor. Camera exposure is set to 30 ms with an EM gain of 300.

6. Some of the fluorescent antibody will attach directly to the lipids in the bilayer through a streptavidin-biotin bond. These are helpful to assess the quality of the lipid bilayer (by checking the mobility) and to find the TIRF angle that best reduces the background noise produced by fluorophores in solution.

7. Take ~20 μL of cells and warm them at 37 °C for 5 min before dropping them onto the bilayer.

8. Begin imaging as soon as the cells start contacting the surface. Bright field and IRM images are taken before the start of time-lapse acquisition. For each of the next 12 min, the acquisition

protocol will take a 1000 frame image series in stream mode with a resulting frame rate of 33 frames per second. After each 1000 frame image series, manually switch to IRM channel and take a snapshot before returning to the time-lapse acquisition.

3.7 Image Analysis

The single-particle tracking technique has become widely used to study movement of molecules in cells. Improvements in experimental approaches (e.g., fluorophores more resistant to photobleaching and more sensitive camera sensors) have facilitated higher resolution data and led to the development of sophisticated algorithms [28] to accurately obtain particle positions and connect consecutive frames to generate trajectories. In general, sub-pixel position accuracy is obtained by fitting a Gaussian function to the detected particle as shown in Fig. 2b. Another common trait across algorithms is the determination of a search radius for the position of the particle in the next frame. Once trajectories are obtained (Fig. 3), the next step is to calculate the mean square displacement (MSD) by calculating the average distance that the particle moved between consecutive frames for different time intervals or lags (Fig. 4). The MSD versus time lag curve provides useful information about the physical interaction between the molecule and its environment. In particular, the diffusion coefficient is an important measure of the properties of the medium in which the particle is moving and of its interactions with other molecules. The slope of the MSD as a function of lag time provides a measure of the lateral diffusion coefficient (D) based on the equation:

$$<r(\Delta t)^2> = 4D\Delta t$$

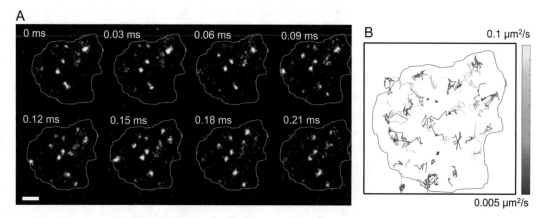

Fig. 3 Single molecule tracks of BCR. (**a**) Snapshots of a typical movie obtained from which single molecules are detected. Scale bar is 1 μm. (**b**) Tracks obtained from the same cell during a single time lapse acquisition. The color code represents the measured diffusivity

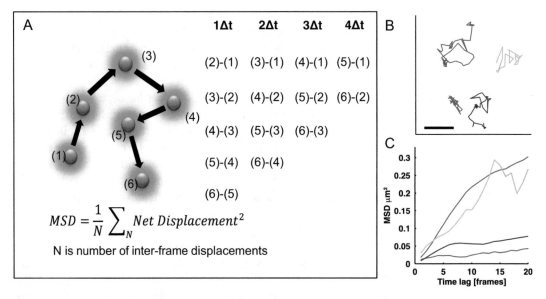

Fig. 4 Mean square displacement (MSD) calculations. (**a**) Procedure to obtain MSD from a set of particle positions over time. (**b**) Four different representative tracks of moving BCR molecules. Scale bar 0.5 μm (**c**) Respective MSDs of the four tracks (color-matched)

4   Notes: Single-Molecule Imaging

1. For well preparation, use caution not to disrupt the bilayer by touching the bottom glass surface or washing wells too quickly with PBS. Wells should remain covered at all times to avoid evaporation and contamination.

2. For cleaning glass slides left overnight in nanostrip solution transfer the slides into a holder filled with 70% ethanol making sure the slides are completely covered. Then wash 20 times with double deionized water and submerge in 70% ethanol again before drying with filtered air.

3. During mouse dissection, wash the left side of the mouse (where the spleen is) with 70% ethanol, to avoid any hair contamination. If hair is on the spleen after removal, wash in a separate petri dish with DMEM before macerating. Remove any fatty tissue associated with spleen to help in B cell isolation.

4. It is important to turn on the stage incubator before imaging to allow temperature stabilization to 37 °C to minimize the drift caused by temperature changes.

In the following sections, we describe the method for traction force microscopy in which cells are activated and allowed to spread on elastic gels which have a layer of beads embedded in them as fiduciary markers (Fig. 5a). As the cells spread, they deform the gel

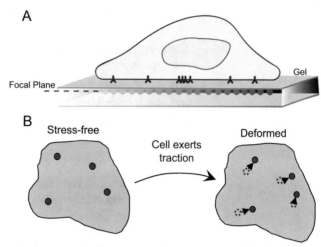

Fig. 5 Schematic illustrating cell-substrate interaction. (**a**) Immune cell activated by an specific antigen (*dark blue*) spreading on an elastic gel (*grey*) with beads (*red*) embedded in the gel. (**b**) Movement of the cell will create stresses on the gel represented by displacement of the embedded beads. Bead displacements are used to calculate to the force exerted by the cell during the immune response

and consequently cause the beads to move from their original position (Fig. 5b). The displacement map of the beads along with the gel stiffness is used to obtain the traction stresses exerted by the cell.

5   Materials: Traction Force Microscopy

5.1   Coverslip Activation

1. Glass bottom dishes of 35 mm diameter with No. 1 coverslip (0.13–0.16 mm thick) (*see* **Note 1**).

2. 2% 3-aminopropyltrimethoxysilane. Add 2 mL of 3-aminopropyltrimethoxysilane to 98 mL of isopropanol inside a chemical hood. Due to reactivity with plastic use a glass pipette and store it in a glass bottle.

3. 1% Glutaraldehyde. Add 1 mL of glutaraldehyde to 99 mL of double deionized water (ddH_2O). Store at 4 °C.

5.2   Polyacrylamide Gels

1. 18 mm circular No. 2 coverslips (0.19–0.23 mm thick).

2. Rainx glass water repellent.

3. 10 mM HEPES (pH: 8.5). Add 1 mL of 1 M HEPES solution to 499 mL of ddH_2O. Adjust pH with sodium hydroxide (NaOH). Adjust volume with ddH_2O to a final volume of 1 L.

4. 2% BIS Solution. Weigh 20 g of BIS powder and mix it with 1000 mL of ddH$_2$O, filter through a 0.45 µm Corning Filter. Transfer to a dark bottle and store at 4 °C. 2% Solution can be bought ready from different companies.

5. 40% Acrylamide Solution. Weigh 400 g of Acrylamide powder and mix it with 1000 mL. Filter and store it as the BIS solution.

6. 10% Ammonium Persulfate in HEPES solution. Discard after use. It cannot be stored.

7. N,N,N,N'-Tetramethyl-ethylenediamine (TEMED).

8. FluoSpheres® Carboxylate-Modified Microspheres. 200 nm red fluorescent beads were used (Excitation 580 nm, Emission 605 nm) (*see* **Note 2**).

5.3 Protein Coating

1. 50 mM Sodium Acetate (pH:4.5). Add 16.66 mL of 3 M sodium acetate to 483.33 mL of ddH$_2$O. Adjust pH with glacial acetic acid. Adjust volume to 1 L with ddH$_2$O.

2. 20× sodium meta-periodate. Mix 80 mg of sodium meta-periodate with 1 mL of 50 mM sodium acetate buffer.

3. Poly-L-lysine sterile-filtered 0.01%.

4. Antibodies for cell activation.

5. Hydrazine Hydrate (undiluted).

6. 5% acetic acid in ddH$_2$O.

6   Methods: Traction Force Microscopy

All the procedures should be carried out at room temperature unless otherwise specified.

6.1 Coverslip Activation

1. Add 500 µL of 2% 3-aminopropyltrimethoxysilane to a glass-bottom dish. Incubate for 15 min (*see* **Note 3**).

2. Wash four times with ddH$_2$O, discarding the solution as chemical waste. Incubate for 10 min in ddH$_2$O.

3. Remove ddH$_2$O and dry at 37 °C for 40 min (*see* **Note 4**).

4. Add 500 µL of 1% Glutaraldehyde. Incubate for 10 min.

5. Wash four times with ddH$_2$O and properly dispose chemical waste. Incubate for 10 min in ddH$_2$O.

6. Remove ddH$_2$O and let it dry at room temperature. Activated dishes can be stored for up to 2 months at room temperature (*see* **Note 6**).

6.2 Polyacrylamide Gel Preparation

Here, we describe a method to prepare elastic substrates consisting of two layers of polyacrylamide gels. The bottom layer is about 50 µm thick, while the top layer has a single layer of fluorescent

Table 1
Mix for three different stiffness values for the first layer of gel (no beads)

Stiffness (Kpa)	Acrylamide (μL)	BIS-Acrylamide (μL)	10% APS (μL)	TEMED (μL)	HEPES (μL)
0.8	281.25	250	25	7.5	4436.25
12	935	500	25	7.5	3532.5
100	1500	1500	25	7.5	1967.5

beads embedded as fiduciary markers and is less than 10 μm thick. The basic polyacrylamide gel preparation technique has been adapted from previously published work [23–25].

1. Prepare stock solutions of acrylamide/bis-acrylamide mix according to Table 1. Solutions can be stored at 4 °C in darkened bottles (*see* **Note 6**).

2. De-gas the solution for 20 min in a vacuum chamber to remove oxygen, which prevents polyacrylamide polymerization.

3. In parallel, prepare 10% Ammonium Persulfate (APS) solution in HEPES buffer.

4. Wipe the 18 mm coverslips vigorously with Rainx using delicate task wipers. Repeat the procedure after drying.

5. Wipe the 18 mm coverslips with ddH$_2$O. The surface of the coverslips is now hydrophobic which improves detachment of the gel after polymerization.

6. Add 7.5 μL of TEMED and 25 μL of 10% APS to the acrylamide solution to initiate polymerization. Mix with a pipette being careful to not introduce bubbles into the mixture (*see* **Note 6**).

7. Add 12.4 μL of solution to the activated bottom glass coverslips for a gel thickness of ~50 μm. Cover the solution gently with the 18 mm treated coverslip on top. Ensure that the solution is spread evenly under the coverslip (*see* **Note 7**).

8. Let the gel polymerize for 40 min at room temperature.

9. Add ddH$_2$O and remove the 18 mm coverslip with the help of tweezers or a razor blade. Wash three times with ddH$_2$O.

10. Prepare solution of acrylamide/bis-acrylamide and carboxylate fluosphere, for the second gel layer, following Table 1 (*see* **Note 8**).

11. Sonicate for 20 min to prevent beads from clustering.

12. Repeat **step 2** to de-gas the solution for the second gel layer.

13. Remove water from the polyacrylamide gels completed in **step 9**. Add 4.1 μL of the acrylamide/beads mix and cover it with 18 mm treated coverslips.

14. Let the gel polymerize at room temperature for 40 min.

15. Add ddH$_2$O and remove the 18 mm coverslip. Wash several times with ddH$_2$O. Store in PBS at 4 °C for up to 4 weeks.

6.3 Coating Gels with Protein

1. Aspirate PBS from gels and add 500 µL of Hydrazine Hydrate. Incubate between 2 and 14 h at room temperature.

2. Dispose Hydrazine Hydrate as chemical waste. Wash the gels three times with ddH$_2$O.

3. Add 500 µL of 5% acetic acid to the gels. Incubate for 1 h.

4. Wash the acetic acid three times with ddH$_2$O. Add 1 mL of ddH$_2$O and incubate for 1 h.

5. In the meantime, in a dark centrifuge tube mix 199.58 µL poly-L-lysine, 99.79 µL of sodium metaperiodate and 1696.46 µL of sodium acetate buffer. Incubate in the dark for 30 min.

6. Remove water from gels and add 130 µL of the protein mix. Incubate gels for 1 h while the crosslinking reaction takes place.

7. Wash the gels several times with PBS. Store at 4 °C in PBS.

8. One day prior to the experiment, remove PBS and add 500 µL of anti-CD3 antibody (we used HIT3A or OKT3 clones), for T cells or monobiotinylated Fab (mbFab, described in previous sections on single-molecule imaging) for B cells. If mbFab is used, first coat gels with streptavidin instead of poly-lysine in **step 5**. Incubate overnight at 4 °C or for 2 h at 37 °C.

6.4 Microscopy

For all imaging experiments, we used a Nikon-TE 2000 PFS microscope with an LED lamp for epi-fluorescence illumination. Images were acquired with an EMCCD camera (Ixon, Andor) or cooled CCD camera (Coolsnap HQ2, Photometrics). For a schematic of the cell on a gel, *see* Fig. 5.

1. Wash gels with PBS three times. Add 1 mL of imaging media. We use the CO$_2$ independent media L-15 (Life Technologies). If the microscope stage is CO$_2$ regulated, then regular cell culture media can be used.

2. After mounting the dish to the microscope, let the temperature equilibrate for at least 15 min to minimize drift during the time-lapse acquisition process.

3. Adjust imaging conditions depending on the experimental design. Select the exposure time of the camera and the imaging rate. We usually set an exposure time of 50 ms. To measure the force generation during T cell and B cell activation, it is sufficient to capture 1 frame every 5 s for 20 min. The displacement field of the beads obtained with this acquisition rate is of sufficient resolution to obtain the force map. If interested in a more detailed examination of a specific step of the immune response, faster acquisition rates can be chosen.

4. Tune the lamp or laser power to obtain a background-beads intensity ratio of at least 5.

5. Explore the gel and focus on a region with a high and uniform density of beads, avoiding areas with clumps of beads.

6. Centrifuge 1 mL of cells from the culture flask for 5 min at $250 \times g$. Resuspend cell pellet in previously warmed imaging media for a final concentration of 2×10^5 cells/mL (*see* **Note 9**).

7. Add 5 µL cells (around 1000) to the gel, being careful not to touch the gel with the pipette. Begin image acquisition immediately after a cell contacts the gel to get the complete regime of immune response. More cells can be added to acquire additional data, ensuring that cells are not too close to each other. Vortex cells gently before adding to the gel to prevent clustering.

6.5 Measuring Gel Stiffness

To measure the stiffness of the polyacrylamide gel, we quantify the stress generated by spheres of different sizes (between 0.3 and 0.5 mm) and masses. The stress is calculated with the indentation generated by the weight of the spheres and the known Poisson ratio of polyacrylamide gels.

1. After the imaging of cells is completed, add a sphere onto the gel close to the area where the cell is located. Using the z-focus of the microscope to record the position of the beads before and after the sphere is added, measure the indentation due to the mass of the sphere. Record the thickness of the gel by recording the z-position of the bead layer and the bottom of the gel at the plane of the coverslip. Calculate the force exerted by the ball: $F = \frac{4}{3}\pi r^3 g(\rho_s - \rho_w)$ where g is the acceleration of gravity, ρ_s is the density of the sphere, and ρ_g is the density of the imaging media.

2. Calculate the Young modulus using the Poisson's ratio for PAA gels ($\nu = 0.48$): $E = \frac{3}{4}(1 - \nu^2)\frac{F}{\sqrt{r \times \varepsilon^3}}$ where ε corresponds to the gel indentation due to the ball.

3. Calculate the correction factor (α) due to the sphericity of the ball: $x = \frac{\sqrt{r\varepsilon}}{h}$ where h is the height of the gel, $\alpha = \frac{1}{1+1.33x+1.203x^2+0.769x^3+0.0975x^4}$.

4. Calculate the corrected Young Modulus for the gel: $\Upsilon = E\alpha$.

6.6 Image Analysis and Force Calculation

1. Store the time-lapse movies acquired as a single-channel tiff file, with a null force image (without cell) as the first frame.

2. Use sub-pixel localization of a few bright beads in cell free regions to account for lateral drift over the time course of the experiment. Shift the position of the beads under the cells by the measured drift at each time point.

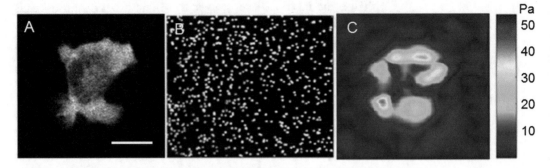

Fig. 6 Traction force measurement. (**a**) EGF-Myosin labeled E6-1 Jurkat T cell on a 0.8 kPa polyacrylamide gel coated with anti-CD3 antibody. Scale bar is 5 μm. (**b**) Single layer of 200 nm red fluorescent beads embedded in the gel. (**c**) Traction stress map showing the stresses exerted by the cell in Pa

3. Perform particle image velocimetry (PIV) to each movie acquired. There are several freely available software packages. For PIV, a pair of consecutive images is divided into smaller regions. The difference between cross-correlation of the region images measures the optical flow (corresponds to velocity or displacement of the objects). The regions can be chosen to be smaller for higher accuracy. We used the MATLAB (Mathworks) based program developed by the Danuser lab [29].

4. Use the PIV data to calculate the traction stress map using Fourier Transform Traction Cytometry (FTTC) [24]. Fourier transform traction cytometry (FTTC) discretizes the moment matrix integral on a grid and takes advantage of the much smaller size of the linear equations set in Fourier Space. It assumes that the gel is a semi-infinite medium. This approximation is valid if displacements are smaller than the thickness of the gel. However, we adapted the FTTC method to account for the finite thickness of the gels [30].

5. The reconstituted force field can be displayed as a heat map of the stress magnitude or as the vector field of the forces (Fig. 6).

7 Notes: Traction Force Microscopy

1. Depending on the needs of the researcher, varied sizes of dishes can be used. All the reactions can be scaled with respect to area without affecting the outcome. Number 1 coverslips are preferred as they allow the use of high numerical aperture oil immersion objectives, which have a small working distance.

2. Spheres smaller than 200 nm do not change the elastic properties of the gel. If higher bead density is required, it is possible to use 40 nm spheres with two different fluorophores to improve spatial resolution.

3. Due to the reactivity of the 3-aminopropyltrimethoxysilane with plastic, square coverslips can be treated with the same protocol. After activation, coverslips can be attached to a circular dish.

4. The drying and manipulation of the coverslips must be done in a sterile and clean environment. Dust particles landing on the glass will decrease the attachment rate of the gel and ruptures can be observed. To minimize this, cover the coverslips with aluminum foil during the drying procedure.

5. To reach the best results, conjugate the activated coverslips with gel the same day.

6. After adding the catalyst for polymerization continue with the protocol as fast as possible. Polymerization will begin immediately and any delays can affect the final mechanical structure of the gel. Temperature affects polymerization rate and therefore the typical length will be affected by it, be sure that the mix is at room temperature before adding the catalysts. TEMED can be stored for a maximum of 1 year without efficiency reduction.

7. Thickness of the gel can be varied to a minimum of 10 μm at which thickness boundary conditions become important and the effect of the glass bottom on the gel stiffness can no longer be ignored.

8. If there is access to a confocal microscope, two layer gels are not needed. Follow Table 1 to make gels with a single layer and focus the microscope to the topmost layer of beads. Add beads to a final concentration of 5% to the mixture, the concentration can be increased to increase beads density.

9. Cells should be subcultured the day prior to the experiments and cell density in the flask should not be higher than 6×10^5 cells/mL.

Acknowledgments

We thank Dr. Thomas Blanpied and Dr. Peter Luo (University of Maryland, School of Medicine) for their help and guidance in establishing the experimental protocol, single-molecule localization, and tracking. We thank Dr. King Lam Hui for optimizing the protocol for making the 2-layer gel used for traction force microscopy. This work was supported by the grants NIH AI122205 and NSF 1563355 to AU.

References

1. Bonilla FA, Oettgen HC (2010) Adaptive immunity. J Allergy Clin Immunol 125 (2 Suppl 2):S33–S40

2. Dustin ML, Groves JT (2012) Receptor signaling clusters in the immune synapse. Annu Rev Biophys 41:543–556

3. Harwood NE, Batista FD (2010) Early events in B cell activation. Annu Rev Immunol 28:185–210

4. Fleire SJ et al (2006) B cell ligand discrimination through a spreading and contraction response. Science 312(5774):738–741

5. Song W et al (2013) Actin-mediated feedback loops in B-cell receptor signaling. Immunol Rev 256(1):177–189

6. Ketchum C, Miller H, Song W, Upadhyaya A (2014) Ligand mobility regulates B cell receptor clustering and signaling activation. Biophys J 106(1):26–36

7. Treanor B, Batista FD (2010) Organisation and dynamics of antigen receptors: implications for lymphocyte signalling. Curr Opin Immunol 22(3):299–307

8. Treanor B, Depoil D, Bruckbauer A, Batista FD (2011) Dynamic cortical actin remodeling by ERM proteins controls BCR microcluster organization and integrity. J Exp Med 208(5):1055–1068

9. Treanor B, Harwood NE, Batista FD (2009) Microsignalosomes: spatially resolved receptor signalling. Biochem Soc Trans 37(Pt 5):1014–1018

10. Batista FD, Treanor B, Harwood NE (2010) Visualizing a role for the actin cytoskeleton in the regulation of B-cell activation. Immunol Rev 237(1):191–204

11. Liu C et al (2013) N-wasp is essential for the negative regulation of B cell receptor signaling. PLoS Biol 11(11):e1001704

12. Liu C et al (2011) A balance of Bruton's tyrosine kinase and SHIP activation regulates B cell receptor cluster formation by controlling actin remodeling. J Immunol 187(1):230–239

13. Wan Z et al (2013) B cell activation is regulated by the stiffness properties of the substrate presenting the antigens. J Immunol 190(9):4661–4675

14. Giannone G et al (2010) Dynamic superresolution imaging of endogenous proteins on living cells at ultra-high density. Biophys J 99(4):1303–1310

15. Betzig E et al (2006) Imaging intracellular fluorescent proteins at nanometer resolution. Science 313(5793):1642–1645

16. Manley S et al (2008) High-density mapping of single-molecule trajectories with photoactivated localization microscopy. Nat Methods 5(2):155–157

17. Bates M, Huang B, Dempsey GT, Zhuang X (2007) Multicolor super-resolution imaging with photo-switchable fluorescent probes. Science 317(5845):1749–1753

18. Rust MJ, Bates M, Zhuang X (2006) Sub-diffraction-limit imaging by stochastic optical reconstruction microscopy (STORM). Nat Methods 3(10):793–795

19. Thompson RE, Larson DR, Webb WW (2002) Precise nanometer localization analysis for individual fluorescent probes. Biophys J 82(5):2775–2783

20. Murase K et al (2004) Ultrafine membrane compartments for molecular diffusion as revealed by single molecule techniques. Biophys J 86(6):4075–4093

21. Natkanski E et al (2013) B cells use mechanical energy to discriminate antigen affinities. Science 340(6140):1587–1590

22. Hui KL, Balagopalan L, Samelson LE, Upadhyaya A (2015) Cytoskeletal forces during signaling activation in Jurkat T-cells. Mol Biol Cell 26(4):685–695

23. Dembo M, Wang YL (1999) Stresses at the cell-to-substrate interface during locomotion of fibroblasts. Biophys J 76(4):2307–2316

24. Butler JP, Tolic-Norrelykke IM, Fabry B, Fredberg JJ (2002) Traction fields, moments, and strain energy that cells exert on their surroundings. Am J Physiol 282(3):C595–C605

25. Gardel ML et al (2008) Traction stress in focal adhesions correlates biphasically with actin retrograde flow speed. J Cell Biol 183(6):999–1005

26. Peluso P et al (2003) Optimizing antibody immobilization strategies for the construction of protein microarrays. Anal Biochem 312(2):113–124

27. Li TP, Song Y, MacGillavry HD, Blanpied TA, Raghavachari S (2016) Protein crowding within the postsynaptic density can impede the escape of membrane proteins. J Neurosci 36(15):4276–4295

28. Chenouard N et al (2014) Objective comparison of particle tracking methods. Nat Methods 11(3):281–289

29. Han SJ, Oak Y, Groisman A, Danuser G (2015) Traction microscopy to identify force modulation in subresolution adhesions. Nat Methods 12(7):653–656

30. Del Alamo JC et al (2007) Spatio-temporal analysis of eukaryotic cell motility by improved force cytometry. Proc Natl Acad Sci U S A 104(33):13343–13348

31. Sohn HW, Tolar P, Pierce SK (2008) Membrane heterogeneities in the formation of B cell receptor–Lyn kinase microclusters and the immune synapse. J Cell Biol 182:367 LP-379

Chapter 5

DNA-Based Probes for Measuring Mechanical Forces in Cell-Cell Contacts: Application to B Cell Antigen Extraction from Immune Synapses

Katelyn M. Spillane and Pavel Tolar

Abstract

The production of antibodies requires the expansion and selection of high-affinity B cell clones. This process is initiated by antigen uptake through the B cell receptor (BCR), which recognizes and binds antigen displayed on the surface of an antigen-presenting cell (APC). To acquire the antigen, B cells use myosin contractility to physically pull BCR-antigen clusters from the APC membrane. These mechanical forces influence association and dissociation rates of BCR-antigen bonds, resulting in affinity-dependent acquisition of antigen by B cells. Mechanical regulation of B cell antigen acquisition from APCs remains poorly understood, although the recent development of DNA-based force sensors has enabled the measurement of mechanical forces generated in B cell-APC contacts. In this chapter, we describe a protocol to design, synthesize, and purify DNA-based force sensors to measure B cell antigen extraction forces using fluorescence microscopy.

Key words B cells, B cell receptor, Immune synapse, Antigen internalization, DNA force sensor, DNA-protein conjugation, Cell mechanics, Fluorescence microscopy

1 Introduction

Antibody responses require the expansion and selection of high-affinity B cell clones. This process is initiated when the B cell receptor (BCR) binds antigens displayed on the surfaces of antigen-presenting cells (APCs), in a cell-cell contact known as the immune synapse. This binding results in BCR signaling, immune synapse formation, and extraction and internalisation of antigen molecules. B cells process the internalized antigen and present it complexed with major histocompatibility complex class II (MHCII) molecules to helper T cells, which provide signals required for B cell activation [1]. T cells provide more help to high-affinity B cells than low-affinity B cells, suggesting that affinity discrimination of antigens during extraction from APCs is critical for B cell clonal selection.

Chaohong Liu (ed.), *B Cell Receptor Signaling: Methods and Protocols*, Methods in Molecular Biology, vol. 1707,
https://doi.org/10.1007/978-1-4939-7474-0_5, © Springer Science+Business Media, LLC 2018

There is growing evidence that B cell acquisition of antigens from APCs is a mechanical process. During immune synapse formation, the B cell cytoskeleton undergoes dynamic rearrangement, generating forces to gather antigens into microclusters and then pull repetitively on BCR-antigen bonds [2]. Pulling on these bonds is mediated by myosin IIa contractility and generates forces of about 10 pN per antigen molecule [3, 4]. These physical processes influence the association and dissociation rates of BCR-antigen bonds and are a way for B cells to test the quality of antigen binding and adapt to a wide range of BCR-antigen affinities [5]. Specifically, because BCR-antigen bonds are slip bonds, mechanical forces rupture bonds with low-affinity antigens, but not high-affinity antigens. This mechanism preferentially promotes antigen extraction, and therefore a biological response, by high-affinity B cell clones. However, although there is a growing appreciation for the role of mechanical forces in regulating B cell responses, the process remains poorly understood, in part because of a lack of appropriate methods to measure the dynamic forces generated in the immune synapse.

Traditional approaches to measuring mechanical forces in biological processes, such as atomic force microscopy and optical and magnetic tweezers, are powerful tools for investigating single-molecule forces such as those generated by molecular motors or nucleic acid enzymes [6]. However, these methods require exogenous probes that are incompatible with in situ measurements of mechanical forces generated in cell-cell contacts. Methods based on the deformability of polymer substrates and micropillar arrays provide information about traction forces in cellular adhesions, although these approaches measure forces averaged over many bonds and detect lateral rather than tensile forces. Molecular tension probes that report strain across a molecular spring using a Förster resonance energy transfer (FRET) fluorophore pair provide high-resolution, single-molecule force measurements in cellular adhesions and can be genetically encoded [7, 8], although these probes can be difficult to design and calibrate.

To overcome these limitations, several groups have developed DNA-based force sensors that use a fluorophore-quencher or fluorophore-fluorophore pair to provide a digital readout of when a DNA duplex unfolds [4, 9–11]. The force at which unfolding occurs can be tuned from ~4 to 70 pN [10, 12], which is suitable range for measuring many mechanotransduction processes in cells. Importantly, DNA-based sensors can be tethered to artificial substrates of varying stiffness and even live cells [11]. This feature allows investigation of how cells modulate their mechanical responses based upon the physical properties of the substrate.

To measure mechanical forces generated by B cells through the BCR, it is necessary to link the DNA force sensor to a protein antigen that the BCR can bind. Here, we describe a protocol for

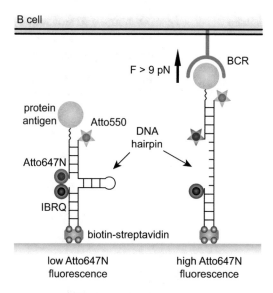

Fig. 1 Schematic of the DNA-based force sensor. The black "ladders" depict DNA. Forces greater than 9 pN applied to the BCR-antigen bond open the DNA hairpin. This opening separates the Atto647N fluorophore from the Iowa Black RQ (IBRQ) dark quencher, resulting in high Atto647N fluorescence

covalently coupling protein antigen to a DNA force sensor that provides a dynamic readout of pulling forces generated by B cells during antigen extraction (Fig. 1). The sensor contains a DNA hairpin with a 9 pN force threshold and a fluorophore-quencher pair that reports high fluorescence when forces above the threshold are generated, and low fluorescence when the force threshold is not reached. The sensors contain a biotin modification so they can be tethered to any substrate through a streptavidin linker. The conjugation approach to linking the protein to the DNA sensor is general and can be used to link any protein that has surface lysine residues to any thiol-modified DNA sensor. We envisage that sensors of similar design can be used to study dynamic, mechanical processes in a wide variety of cell-cell contacts, including T cell receptor-peptide MHC clusters in T cell immune synapses, and focal adhesions in adherent or migrating cells.

2   Materials

All buffer solutions should be prepared using ultrapure water and filtered through a 0.22 μm filter. Proteins of the highest purity should be used. DNA oligomers should be purchased with HPLC purification. Degas the conjugation, start, and elution buffers under vacuum for 1 h before use.

2.1 DNA Oligomer Sequences

1. 9 pN hairpin strand sequence 5′-C6 NH$_2$-tca cga cag gtt cct tcg cat cga tat tat ata tta ata tat aa tttt tta tat att aat ata taa ta att tac tca caa gca gtg tgt aca-biotin.

2. Upper handle complementary strand sequence 5′-Atto647N-tcg atg cga agg aac ctg tcg tga-S-S.

3. Lower handle complementary strand sequence 5′-biotin-tgt aca cac tgc ttg tga gta aat-Iowa Black RQ.

2.2 DNA-Protein Conjugation

1. Conjugation buffer: PBS, pH 7.3, 1 mM EDTA.

2. Duplex buffer: 30 mM HEPES, pH 7.5, 100 mM potassium acetate.

3. Single-stranded DNA oligomers purified by HPLC, resuspended to 100 μm in duplex buffer (store at −20 °C) (*see* **Note 1**).

4. ChromPure goat IgG, F(ab′)$_2$ fragment (Jackson ImmunoResearch).

5. Zeba spin desalting columns, 7k MWCO, 0.5 mL (Thermo).

6. Vivaspin 500 centrifugal concentrators, 10k MWCO, polyethersulfone membrane.

7. NIP-Osu (Biosearch Technologies).

8. SMCC, water-soluble (sulfo-SMCC; Calbiochem), dissolved to 100 mM (43.6 mg/mL) in conjugation buffer immediately before use.

9. Dithiothreitol (DTT; 1 M in conjugation buffer), made fresh immediately before use.

10. Labeling buffer: 0.1 M NaHCO$_3$, pH 8.3.

11. Atto550 NHS ester.

12. Anhydrous dimethyl formamide (DMF).

13. Magnesium chloride (MgCl$_2$; 100 mM in H$_2$O).

14. 100 bp DNA ladder (New England Biolabs).

2.3 Conjugate Purification

1. Ultrapure water (Milli Q), degassed.

2. Start buffer: 25 mM Tris, pH 8 at 4 °C, 100 mM NaCl (store at 4 °C), degassed.

3. Elution buffer: 25 mM Tris, pH 8 at 4 °C, 1000 mM NaCl (store at 4 °C), degassed.

2.4 Equipment

1. Mono Q 5/50 GL column (GE Healthcare).

2. AKTA protein purification system.

3. NanoDrop.

4. Variable-speed benchtop microcentrifuge.

5. UV transilluminator (Gel Doc EZ gel documentation system, Bio-Rad, or similar).

6. E-Gel general purpose agarose gels, 2% (Invitrogen).

7. E-Gel PowerBase (Invitrogen).

8. Thermal cycler.

9. ThermoMixer.

3   Methods

This protocol describes the design and stoichiometric conjugation (1:1) of a DNA-based dynamic force sensor to a goat IgG F(ab')$_2$ protein fragment (Fig. 2). After annealing, but prior to conjugation, the NH$_2$ group on the DNA hairpin strand is labeled with Atto550 NHS ester to allow ratiometric imaging using Atto550 and Atto647N fluorescence. The NH$_2$ groups of the protein lysine residues are activated with maleimide using the heterobifunctional cross linker SMCC, and the DNA sensor thiol group is reduced with DTT to make it reactive toward maleimides. The SMCC-activated protein and thiol-reduced DNA sensor are then mixed together in a 5:1 protein:DNA molar ratio. The excess protein in the reaction ensures that each protein is conjugated to only one DNA sensor. Following purification to remove unreacted protein and DNA, the conjugated protein is haptenated with 4-hydroxy-3-iodo-5-nitrophenylacetic acid (NIP) to allow stimulation of NIP-specific B1–8 mouse primary B cells. The conjugation procedure has been followed successfully using several different DNA-based molecular sensors (mechanism, tension, and dynamic force sensors) [4, 11] as well as proteins (goat IgG F(ab')$_2$, anti-human IgM F(ab')$_2$, goat anti-mouse Igκ F(ab')$_2$, hen egg white lysozyme). Possible post-purification modifications to the conjugated protein include haptenation and/or fluorescent labeling. The procedure can be spread over 2 days, with the conjugation reaction performed on day 1 and the purification and post-purification modifications on day 2, without any noticeable degradation of the sample.

3.1   DNA-Protein Conjugation

1. In a PCR tube, mix 10 μL of each 100 μm single-stranded DNA oligomer and 0.61 μL of 100 mM MgCl$_2$ (final MgCl$_2$ concentration of 2 mM). Anneal the mixed oligonucleotides by heating to 94 °C for 2 min in a thermocycler and transferring the PCR tube to the benchtop for 30 min to gradually cool to room temperature (*see* **Note 2**).

2. Equilibrate one Zeba spin column with labeling buffer following the manufacturer's instructions (*see* **Note 3**).

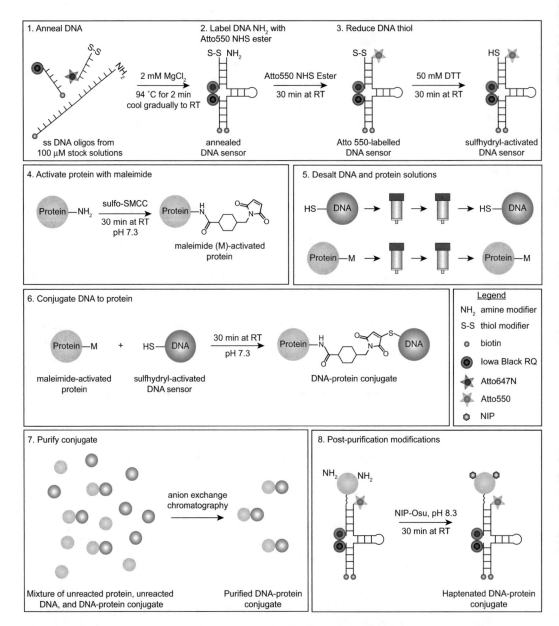

Fig. 2 Workflow for producing DNA-protein conjugates. Single-stranded DNA oligomers are annealed to form the DNA hairpin force sensor. The primary amine group on the sensor is labeled with Atto550 NHS ester and the thiol converted to a reactive sulfhydryl through incubation with DTT. The heterobifunctional linker sulfo-SMCC is used to activate protein primary amines with maleimide groups. Following a desalting step to exchange the DNA and protein into labeling buffer, the DNA and protein are mixed together to allow conjugation of the DNA sulfhydryl with the protein maleimide. The DNA-protein conjugates are purified from unreacted DNA and protein by anion exchange chromatography. Further modifications, such as fluorophore labeling or haptenation, can be done following this purification step

3. Spin the annealed DNA sensor through the Zeba column to exchange the DNA into labeling buffer.

4. Add 0.48 µL of 5 mg/mL (6.32 mM) Atto550 NHS ester (in anhydrous DMF) to the DNA solution and mix well by pipetting to label the hairpin strand NH_2 functional group with Atto550. Incubate at room temperature, for 30 min, with mixing using a ThermoMixer.

5. During the incubation in **step 4**, concentrate 120 µL of 4.7 mg/mL goat IgG F(ab′)$_2$ to 40 µL by spinning at 15,000 × g for 3–4 min in a Vivaspin 500 centrifugal concentrator (*see* **Note 4**).

6. Equilibrate two Zeba spin columns with conjugation buffer following the manufacturer's instructions.

7. Spin the Atto550-labeled DNA sensor through one Zeba column to remove unreacted dye and exchange into conjugation buffer. Spin the concentrated protein solution through the second Zeba column to exchange into conjugation buffer.

8. Make a fresh 1 M (154.25 mg/mL) solution of DTT in conjugation buffer and add 1.5 µL to the DNA sensor to reduce the DNA thiol (final DTT concentration of 50 mM). Incubate at room temperature, for 30 min, with mixing using a ThermoMixer.

9. Dissolve a small amount of SMCC in conjugation buffer to make a 10.9 mg/mL (25 mM) solution (*see* **Note 5**).

10. Add 1 µL of the SMCC solution to the 40 µL protein solution and mix well. This results in a ~5-fold molar ratio of SMCC to protein. Incubate at room temperature, for 30 min, with mixing using a ThermoMixer.

11. Equilibrate four Zeba spin columns with start buffer following the manufacturer's instructions.

12. Spin both the reduced DNA sensor and the SMCC-labeled protein through two Zeba columns sequentially to remove DTT and excess SMCC.

13. Combine the DNA and protein solutions in the same tube and mix well by pipetting. Incubate at room temperature, for 1 h, with mixing using a ThermoMixer.

14. Load 1 µL of the DNA-protein mixture onto a 2% agarose gel along with a 100 bp DNA ladder to confirm a successful conjugation reaction. The conjugate is identified by a mobility shift on the gel (Fig. 3a and *see* **Note 6**).

3.2 Conjugate Purification

Typically, 35–80% of the DNA sensor is coupled to the protein during the above conjugation reaction, as determined by agarose gel electrophoresis (Fig. 3a). The unreacted DNA and protein in

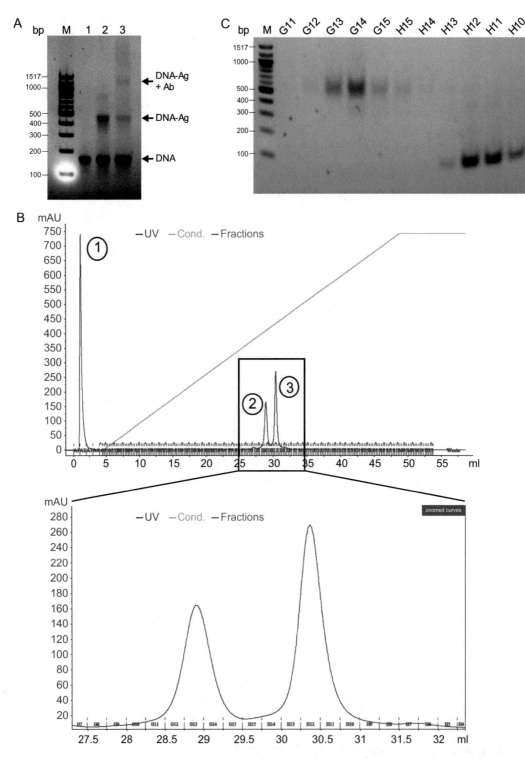

Fig. 3 Synthesis and purification of DNA-protein conjugates. (**a**) Agarose gel electrophoresis (2% agarose) showing the 100-bp marker (*M*), annealed DNA sensor (*lane 1*), DNA-protein conjugate (*lane 2*), and the DNA-protein conjugate mixed with a secondary antibody against the protein (*lane 3*). (**b**) Anion exchange

the reaction mixture are removed using a Mono Q anion exchange column (Fig. 3b). The presence of purified DNA-protein conjugates is confirmed by agarose gel electrophoresis of the DNA-containing fractions (Fig. 3c).

1. Equilibrate the Mono Q column with start buffer following the manufacturer's instructions.

2. Apply the DNA-protein mixture to the column and wash the column with 5 column volumes of start buffer.

3. Elute the DNA-protein conjugate using a gradient of 0–100% elution buffer over 45 column volumes, at a flow rate of 0.5 mL/min, collecting 0.25 mL fractions. The first peak to elute is the unconjugated protein, the second is the DNA-protein conjugate, and the third is the unconjugated DNA (Fig. 3b, c).

4. Concentrate DNA-protein conjugate fractions by spinning at 15,000 × g for 3 min in Vivaspin 500 centrifugal concentrators.

5. Load 4 μL of each concentrated fraction onto a 2% agarose gel along with a 100 bp DNA ladder to identify fractions containing the purified DNA-protein conjugate.

6. Pool fractions containing the purified DNA-protein conjugate and measure the concentration using a NanoDrop (*see* **Note 7**).

3.3 Post-Purification Modifications

Following purification, the conjugated protein is haptenated with a 50-fold excess of NIP.

1. Equilibrate a Zeba spin column with labeling buffer following the manufacturer's instructions.

2. Spin the purified DNA-protein conjugate through the Zeba column to exchange into labeling buffer.

3. Dissolve a small amount of NIP-Osu into labeling buffer to make a 4.2 mg/mL (10 mM) solution.

4. Add a 50-fold molar excess of NIP-Osu to the DNA-protein conjugate solution and mix well. Incubate at room temperature, for 1 h, with mixing using a ThermoMixer.

Fig. 3 (continued) (Mono Q) UV elution profile (*blue*) of the DNA-protein conjugate. The fractions (*red*) are eluted over 45 column volumes with a NaCl gradient of 100–1000 mM (*green*). The first peak to elute is unconjugated protein, the second peak is the DNA-protein conjugate, and the third peak is the unconjugated DNA. (**c**) Agarose gel electrophoresis (2% agarose) of the DNA-containing fractions (*G11–H10*) from the elution profile in (**b**). Note that **a** and **b/c** represent different preparations of DNA-protein conjugates, but are representative of all preparations. ©SPILLANE et al., 2017. Originally published in THE JOURNAL OF CELL BIOLOGY. https:/doi.org/10.1083/jcb.201607064 [11]

5. Equilibrate a Zeba spin column with PBS following the manufacturer's instructions.

6. Spin the DNA-protein conjugate through the Zeba column to remove unreacted NIP-Osu.

7. Measure the concentration of DNA-protein conjugate and NIP-Osu using a NanoDrop to determine the degree of labeling.

8. Freeze the sample in 2 μL aliquots at −20 °C (*see* **Note 8**).

4   Notes

1. Single-stranded DNA oligomers should be resuspended at 100 μM in duplex buffer, stored at −20 °C, and thawed as needed. Single-stranded oligomers are very stable over many freeze/thaw cycles, and are more stable in buffer than in water.

2. To scale up the size of the conjugation reaction described here, anneal several 30 μL aliquots of DNA sensor in parallel rather than increasing the volume in a single PCR tube. Once the sensors are annealed, the contents of the PCR tubes can be combined into a single tube before exchanging into labeling buffer and labeling with Atto550 NHS ester. The amount of protein used in the conjugation reaction should be scaled up equivalently. The sample volume capacity of 0.5 mL Zeba columns is 130 μL. If the volumes for desalting and buffer exchange exceed 130 μL, Zeba columns with larger capacity should be used (see the manufacturer's instructions to determine the appropriate size column). We recommend always doing a small-scale reaction as described in this protocol when working with a new DNA sensor or protein before attempting a larger-scale preparation.

3. Buffers containing primary amines, such as Tris, should not be used during the DNA-protein conjugation or protein labeling steps because these compete for reaction with the NHS ester reagents.

4. The DNA-protein conjugation yield depends critically on the purity of the reagents. DNA should be purchased with HPLC purification. It may be necessary to purify the protein using gel filtration or anion exchange chromatography prior to using it in the conjugation reaction.

5. Sulfo-SMCC should be stored under argon and equilibrated to room temperature before opening. NHS esters degrade by hydrolysis, which competes with their reaction for primary amines. Therefore, fresh solutions should be prepared in degassed conjugation buffer immediately before use. Once the conjugation buffer is added to the sulfo-SMCC, vortex

briefly (~10 s) to dissolve and add to the protein as quickly as possible. The NHS ester (amine-targeted) reaction should always be done before the maleimide (sulfhydryl-targeted) reaction. In our experience, sulfo-SMCC purchased in no-weigh format (2 mg, single-use aliquots) is not as reactive as that purchased in a 50 mg bottle.

6. If the reaction yield is low (<20%), the molar excess of protein in the reaction may need to be increased, the reactants concentrated, or the reagents purified. The protein should always be reacted in at least five-fold molar excess relative to the annealed DNA sensor. Reagents should be concentrated to at least 25 μM annealed DNA sensor and 100 μM protein (equivalent to 11 mg/mL of an antibody F(ab′)₂ fragment) if possible. It also may be that the sulfo-SMCC crosslinker has degraded, in which case a new bottle should be purchased.

7. Due to the significant overlap of DNA and protein absorbance spectra, the concentration of the DNA-protein conjugate is most easily determined by measuring the concentration of Atto647N. Because each conjugate is designed to have a single Atto647N modification, the concentration of Atto647N is equivalent to the concentration of the DNA-protein conjugate. The concentration is determined using the Beer-Lambert law and the Atto647N extinction coefficient of 150,000/M/cm at the absorbance maximum 647 nm.

8. The conjugation and purification steps can be spread over 2 days without any loss of DNA-protein conjugate stability. In this case, the conjugation should be completed on day 1, and the purification, modification, and aliquoting steps completed on day 2. Aliquots should be thawed and used as needed; do not refreeze or store at 4 °C for more than 1 day.

References

1. Batista FD, Iber D, Neuberger MS (2001) B cells acquire antigen from target cells after synapse formation. Nature 411:489–494

2. Fleire SJ, Goldman JP, Carrasco YR, Weber M, Batista FD (2006) B cell ligand discrimination through a spreading and contraction response. Science 312:738–741

3. Natkanski E, Lee W-Y, Mistry B, Casal A, Molloy JE, Tolar P (2013) B cells use mechanical energy to discriminate antigen affinities. Science 340:1587–1590

4. Nowosad CR, Spillane KM, Tolar P (2016) Germinal center B cells recognize antigen through a specialized immune synapse architecture. Nat Immunol 17:870–877

5. Tolar P, Spillane KM (2014) Force generation in B-cell synapses: mechanisms coupling B-cell receptor binding to antigen internalization and affinity discrimination. Adv Immunol 123:69–100

6. Neuman KC, Nagy A (2008) Single-molecule force spectroscopy: optical tweezers, magnetic tweezers and atomic force microscopy. Nat Methods 5:491–505

7. Morimatsu M, Mekhdjian AH, Adhikari AS, Dunn AR (2013) Molecular tension sensors report forces generated by single integrin molecules in living cells. Nano Lett 13:3985–3989

8. Grashoff C, Hoffman BC, Brenner MD, Zhou R, Parsons M, Yang MT, McLean MA, Sligar SG, Chen CS, Ha T, Schwartz MA (2010) Measuring mechanical tension across vinculin reveals regulation of focal adhesion dynamics. Nature 466:263–266

9. Blakely BL, Dumelin CE, Trappmann B, McGregor LM, Choi CK, Anthony PC, Duesterberg VK, Baker BM, Block SM, Liu DR, Chen CS (2014) A DNA-based molecular probe for optically reporting cellular traction forces. Nat Methods 11:1229–1232

10. Zhang Y, Ge C, Zhu C, Salaita K (2014) DNA-based digital tension probes reveal integrin forces during early cell adnhesion. Nat Commun 5:5167

11. Spillane KM, Tolar P (2017) B cell antigen extraction is regulated by physical properties of antigen-presenting cells. J Cell Biol 216:217–230

12. Wang X, Ha T (2013) Defining single molecular forces required to activate integrin and Notch signalling. Science 340:991–994

Chapter 6

Deriving Quantitative Cell Biological Information from Dye-Dilution Lymphocyte Proliferation Experiments

Koushik Roy, Maxim Nikolaievich Shokhirev, Simon Mitchell, and Alexander Hoffmann

Abstract

The dye-dilution assay is a powerful tool to study lymphocyte expansion dynamics. By combining time course dye-dilution experiments with computational analysis, quantitative information about cell biological parameters, such as percentage of cells dividing, time of division, and time of death, can be produced. Here, we describe the method to generate quantitative cell biological insights from dye-dilution experiments. We describe experimental methods for generating dye-dilution data with murine lymphocytes and then describe the computational data analysis workflow using a recently developed software package called FlowMax. The aim is to interpret the dye-dilution data quantitatively and objectively, such that cell biological parameters can be reported with an appropriate measure of confidence, which in turn depends on the quality and quantity of available data.

Key words Lymphocyte, B cell, Proliferation, Lymphocyte dynamics, CFSE assay, Quantitative dye dilution, Cell biological parameter, FlowMax

1 Introduction

Lymphocytes expansion dynamics are produced by the division and death of individual lymphocytes within the population. The effective immune response relies on the balance of division and death cell fate decisions. Classical methods to study lymphocyte proliferation are the incorporation of tritiated thymidine [1] or colorimetric assay [2] at a single time point. These methods produce relative information of proliferation but do not produce a deep understanding of the proliferative response under different stimulation conditions. For example, less incorporation of tritiated thymidine in condition X compared to condition Y suggests the following possibilities: (1) a smaller number of cells enter the division cycle in condition X than Y, (2) cells are dividing slower in condition X than Y, (3) cells are dying faster in condition X than Y, (4) the time to the first division is longer in condition X than Y but division times of the

Chaohong Liu (ed.), *B Cell Receptor Signaling: Methods and Protocols*, Methods in Molecular Biology, vol. 1707, https://doi.org/10.1007/978-1-4939-7474-0_6, © Springer Science+Business Media, LLC 2018

following generations are the same, or (5) division rates are a little higher but death rates are much higher in condition X than Y.

These possibilities can be distinguished by time course dye-dilution experiments when followed by quantitative interpretation of dye-dilution data [3]. These dyes bind covalently to intracellular molecules and are fluorescent in nature. Cell divisions are measured by halving of the fluorescence in each division cycle and each peak in a log-fluorescence histogram represents corresponding generation number [4]. A number of theoretical models have been developed to interpret dye-dilution data. However, they do not provide information about the quality of fit to the experimental data [5, 6]. Recently, our laboratory developed an integrated computational tool "FlowMax," which derives cell biological parameters from dye dilution data, provides information about the quality of fit, and thus allows objective interpretation of dye-dilution data to improve the rigor of dye dilution experiments. The FlowMax tool has a graphical user interface, provides the necessary functionality for preprocessing of raw fluorescence data, and is freely available upon request.

Here, we provide protocols for the experimental method of labeling lymphocyte and the computational method for quantitatively analyzing the data. Briefly, B cells are purified from splenocytes by negative selection (CD43 (Ly-48) MicroBeads) using magnetic assisted cell sorting. B cells are labeled with CellTrace™ Far Red (CTR). CTR labeled B cells are stimulated with CpG. 100 μL of culture volumes are acquired by flow cytometry at 14, 36, 48, 72, 96, and 120 h. Dead cells are excluded using dead cell marker (7AAD). The acquired data are exported as a FCS file and imported into "FlowMax." The generation "0" peak is identified on the histogram and a broad range of the cell biological parameters are derived computationally.

2   Material

All reagents should be cell culture grade.

2.1 Isolation of B Cells

1. Phosphate buffer saline (PBS, pH 7.4): 137 mM NaCl, 2.7 mM KCl, 4.3 mM Na_2HPO_4, 1.47 mM KH_2PO_4.

2. Media: 1640 RPMI, 10% FBS, 1× pen-strep, 5 mM glutamine, 1 mM sodium pyruvate, 1 mM MEM non-essential amino acid, 20 mM HEPES, and 55 μM 2-mercaptoethanol (*see* **Note 1**).

3. Room temperature (RT) PBS (*see* **Note 2**).

4. RT media (*see* **Note 3**).

5. 1.5 mL polypropylene centrifuge tube.

6. Table top centrifuge.

7. Cold FBS.

8. Cell Counter.

9. CD43 (Ly-48) Microbeads (Miltenyi Biotec GmbH, 130-049-801) mouse.

10. Red blood cell (RBC) lysis buffer (eBioscience, 00-4333-57).

11. Magnetic assisted cell sorting (MACS) buffer: PBS containing 0.5% BSA with 2 mM EDTA.

2.2 Labelling of Lymphocytes

1. 37 °C PBS.

2. RT media.

3. 1.5 mL polypropylene centrifuge tube.

4. Rotator for mixing.

5. 37 °C incubator.

6. Table top centrifuge.

7. Cold FBS.

8. CellTrace™ Far Red (CTR) Cell Proliferation Kit (Thermo Fisher Scientific, C34572). Add 20 µL of DMSO to one vial of CellTrace™ Far Red. Mix it by either mild vortexing or pipetting up-and-down, followed by a short spin (*see* **Note 4**).

9. Cell counter.

2.3 Proliferation and Stimulation

1. Prepare 2.5×10^5 CTR label cells/mL in media.

2. Stimulus (CpG).

3. 48-well tissue culture plate.

4. Dead cell marker (7AAD, 7-Aminoacetinomycin D).

5. Flow cytometry (Accuri C6).

2.4 FlowMax Analysis

1. PC with minimum configuration i3 processor, 4 GB RAM, 100 GB hard disk and either Linux, MacOSX, or Windows 7/10.

2. Install latest version of Java (https://java.com/en/download/).

3. Download FlowMax as a standalone JAR file

   (http://signalingsystems.ucla.edu/models-and-code/).

3 Methods

3.1 Isolation of B Cells

1. Isolate spleen from 10 to 12 weeks old C57BL/6 mice (*see* **Note 5**).

2. Immediately keep the spleen in cold media on ice.

3. Isolate splenocytes by macerate using strainer and plunger.

4. Centrifuge at 450 rcf, 4 °C for 5 min and discard the media.

5. Resuspend the pellet in 5 mL RBC lysis buffer (*see* **Notes 6 and 7**) and keep in RT for 5 min.

6. Add 5 mL of RT PBS and centrifuge at 450 rcf, 25 °C for 5 min (*see* **Note 8**).

7. Resuspend the cell pellet in 10 mL MACS buffer and count the splenocytes.

8. Centrifuge at 450 rcf, 4 °C for 5 min and discard the media.

9. Resuspend 10^7 splenocytes in 90 µL cold MACS buffer and add 10 µL CD43 (Ly-48) MicroBeads mouse (*see* **Note 9**).

10. Incubate for 15 min in ice or cold chamber (4 °C) with continuous shaking.

11. Adjust the volume to 10 mL and centrifuge at 450 rcf, 4 °C for 5 min.

12. Equilibrate the LS column with 3 mL cold MACS buffer at the time of centrifugation.

13. Discard the MACS buffer after centrifugation (**step 11**) and resuspend in 500 µL MACS buffer for 10^8 splenocytes in **step 7** (*see* **Note 10**).

14. Pass it through the LS column and collect in a 15 mL polypropylene centrifuge tube. Wash the column three times with 1 mL cold MACS buffer.

15. Centrifuge the flow through at 450 rcf, 4 °C for 5 min.

16. Resuspend the pellet in media and count the cell.

17. Verify the purity of the B cell by B220-FITC and 7AAD [7].

3.2 Labelling of Lymphocytes

1. Prepare 1 mM CellTrace™ Far Red (CTR) dye stock solution in DMSO. Aliquot 5 µL of stock solution in a 500 µL polypropylene centrifuge tube and store it in −80 °C (*see* **Note 11**).

2. A. Thaw 1 mM CTR stock at RT and prepare 2 µM working concentration by diluting in warm PBS (37 °C) (*see* **Note 12**). The volume of working concentration is 500 µL.

   B. Prepare 10×10^6 cells/mL of working concentration in warm PBS (37 °C). Make single-cell suspension by pipetting up and down.

3. Add 500 µL of cells (**step 2B**) to 500 µL of CTR solution (**step 2A**) in a 1.5 mL polypropylene centrifuge tube. Mix immediately by inverting the tube 3–4 times. If necessary, mildly vortex the tube for 15–30 s. Final concentration of CTR and cells should be 1 µM and 5×10^6 cells/mL respectively (*see* **Note 13**).

4. Incubate the cells at 37 °C for 20 min with constant mixing. Alternatively, incubate the cells at RT for 25 min with constant mixing.

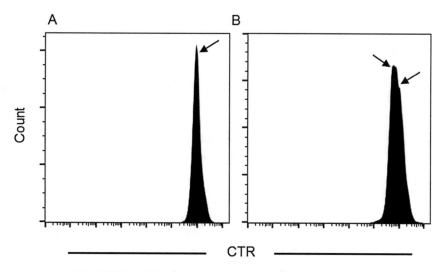

Fig. 1 Fluorescence profile of CTR label B cells. Peak representing CTR label B cells at "0" h is indicated by an *arrow*. (**a**) Single peak with narrow distribution, (**b**) two peaks with wide distribution

5. Quench the excess unreacted CTR by adding 500 μL (1 volume) cold FBS and mix well by inverting the tube.

6. Pellet the cells by centrifuge at 450 rcf, 4 °C for 5 min and resuspend the cells pellet in 1 mL pre-warm (37 °C) media.

7. Repeat **step 6**.

8. Keep the cells 10 min in pre-warm media. Pellet and resuspend the cells as described in **step 6**.

9. Check the efficiency and quality of CTR labeling by flow cytometry (Fig. 1). If the histogram shows multiple peak or wide distribution discard the labeled cells (Fig. 1b) and start to label new cells to achieve a log-fluorescence histogram with single peak and narrow distribution (Fig. 1a).

10. Count the cells.

3.3 Stimulation and Proliferation

1. Add 3 mL media to 7.5×10^5 CTR label cells and prepare single-cell suspension. The working concentration of cells $2.5 \times 10^5/mL$.

2. Mix CpG, to a final concentration of 50 nM, to the single-cell suspension. Invert the tubes 3–4 times to thoroughly mix. Seed 250 μL of cells/ well in a 48-well plate and culture the cells in 37 °C with 5% CO_2 in humidified atmosphere for 14, 36, 48, 72, 96, and 120 h. For all the time point seed in duplicate.

3. Gently pipette the cells in the respective well. Transfer the complete contents of each well to a new 1.5 mL polypropylene centrifuge tube. Add 10 μL of 7AAD (*see* **Note 14**) to each tube and incubate for 5 min. Pass the cells through a 40 μm strainer. Acquire 100 μL (*see* **Note 15**) at each time point using

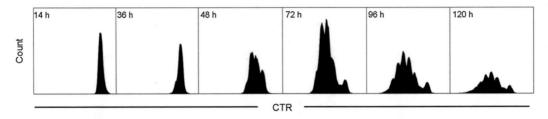

Fig. 2 B cell proliferation time course. CTR label B cells stimulated 50 nM CpG. Dead cells were excluded by staining with 7AAD. B cell proliferation measured at 14, 36, 48, 72, 96, and 120 h

the Accuri C6 flow cytometer (Fig. 2 and *see* **Note 16**). Ensure to independently acquire cells from two replicate wells for each time point. Measure proliferation at 14 h (*see* **Note 17**), 36, 48, 72, 96, and 120 h (*see* **Note 18**).

4. Export the acquired file to Flow cytometry standard (FCS) file (one file for each flow cytometry run, clearly labeled by time from the start of stimulation and replicate number).

3.4 FlowMax

1. FlowMax is written in java 1.6 language. Make sure java is installed on your computer (*see* **Note 19**).

2. Startup FlowMax either by double clicking on the downloaded jar file or by running it from a terminal or command prompt using command "java –jar FlowMax.jar" (*see* **Note 20**).

3. The FlowMax user interface is shown in Fig. 3a. FlowMax has three tabs: "Data," "Phenotyping," and "Solution Analysis." The "Data" tab provides functionality for importing FCS files (raw data), performing in silico compensation, gating on scattering and fluorescence. The data tab also enables annotating each sample according to stimulation time, approximate location of the generation 0 peak, and total cell count (estimated automatically if run on Accuri C6). The "Phenotyping" tab is used for the model fitting of the experimental data using the total simulation time of each sample, and the total cell count information provided by the user on. The "Solution Analysis" tab visualizes the cell biological parameters obtained by fitting the mathematical model of proliferating lymphocytes. The functions of the tabs are discussed below in more detail.

4. From the "Data" tab, click the "Load FCS" button to import the FCS file—it will then appear in the left panel (Fig. 3b). Each time point is shown in a row and double clicking on each row opens the respective FCS file. Fluorescence channel compensation can be performed if needed (*see* Fig. 4 and **Note 21**). Gate on the population of interest (Fig. 3c) based on the Forward scattered (FSC) and Side scatter (SSC). The gated population will appear underneath the sample in the left panel. Double click on the gated population name on the left to display a new

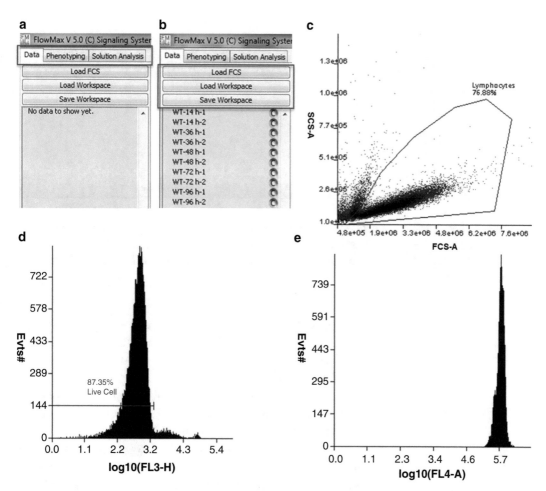

Fig. 3 Using FlowMax to build log-fluorescence histograms. (**a**) FlowMax has tabs for data preprocessing (Data), for model fitting (Phenotyping), and for solution analysis and visualization (Solution Analysis). (**b**) Begin by loading all of the FCS raw datasets, or a previously saved workspace. Datasets and gate information will be listed at left below the buttons. (**c**) Double click a sample to show a plot of the data. Use FCS and SSC to select the viable cell population. (**d**) Use another gate to select the viable cells. (**e**) A log fluorescence histogram showing CTR fluorescence on the log scale. For this time point, cells are still largely undivided

plot. Change the *Y*-axis variable to the viability dye channel (7AAD) by right clicking and selecting this channel from the menu. Now create a gate to distinguish the viable cells from the dead cells (Fig. 3d). From the left panel, double-click the viable cell population. A third plot will appear containing the viable cells. Generate a log fluorescence histogram of CTR by toggling log x, right-clicking on the *X*-axis, and selecting CTR (Fig. 3e).

5. Define generation "0" in the histogram of CTR (*see* **Note 22**), input sample name and condition (cell type, genotype, stimulus), and define the time since stimulation in hours. The sample

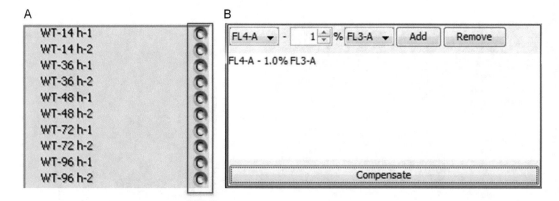

Fig. 4 Manual compensation for fluorescence spillover between channels. (**a**) Click on the multi-colored circle next to a sample to open the compensation dialog box. (**b**) The compensation dialog box can be used to subtract a percentage of one channel from another (in this case 1% of the viability stain (FL3) channel is subtracted from the CTR (FL4) channel)

name must be identical for samples to be associated together. Replicates should be defined as having the same stimulation duration. Save your progress by clicking on "Save Workspace" button in the "Data" tab (Fig. 3b).

6. Repeat **steps 4** and **5** for each collected sample. Multiple conditions/stimulations can be loaded and processed simultaneously but will require additional computational memory.

7. Go to the "Phenotype" tab and select the condition in the left side panel of Fig. 5a. To get a quick estimate of parameters, set "Solutions" to 5, which means that five separate model fitting attempts will be made (*see* **Note 23**). Click "Phenotype FCyton" to start model fitting. Solution number and fitting statistics appear at the bottom of the left side panel. Green line indicates fitting of the entire population, blue line indicates fitting of each generation, and background color indicates the goodness of fit. Green, yellow, and red background of each sample corresponds to <10%, 10–20%, and 20–30% error of the fit respectively (Fig. 5b). For the best model fit, all background colors should be green, and generations should line up with center of peaks. If a model peak appears to be misaligned, or if most samples are fit poorly (red background), please adjust the fluorescence and fcyton parameter ranges (*see* **Note 24**). If satisfied with the preliminary 5 solutions repeat the same run with 500–1000 solution to build a comprehensive sampling of the solution space (*see* **Note 25**).

8. Go to "Solution Analysis" tab to visualize the model solutions (Phenotypes). Phenotypes automatically appear after phenotyping has finished. To visualize previous phenotypes, click on "Load Phenotype," and browse to the saved .csv file (Fig. 6a).

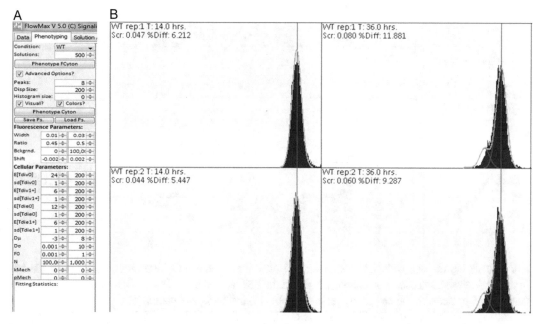

Fig. 5 FlowMax model fitting of CTR labeled B cells. (**a**) The parameter panel in the "Phenotyping" tab is used to set the number of solutions, the parameter ranges, and whether solutions should be visualized while fitting. (**b**) Model fit (*green* and *blue traces*) overlapping the CTR log-fluorescence histograms. The two rows are duplicates and each column indicates a different time point. The condition and time point are shown on the *top left corner* of each box. *Black histogram* is the experimental data. The *green line* is the optimal fit when the data of all generations is considered; the *blue line* is the optimal fit to each generation. The *red line* indicates the fluorescence of generation "0" or undivided cells. *Green* and *yellow backgrounds* indicate the quality of the fits: <10% and <20% deviation from experimental data, respectively. The deviation between experimental data and fit is shown in *blue* at the *top left corner* (below the condition and time point) of the each box

9. Click "Draw" in the tab "Visualize Parameter" to plot cell biological parameters, e.g., Fs (progressor fraction, i.e., % of cells dividing in each generation), Tdiv (Time to division), and Tdie (Time to death). Suffix indicates the cell generation, e.g., F0 means F of generation 0, Tdiv1+ means Tdiv from 1 to terminal generation Parameters on the right are used to change the plot parameters. Plots can be saved by right-clicking and selecting copy to clipboard (*see* **Note 26**). To export the data for plotting in other software use the appropriate save button for the data required at the bottom "visualize parameter" tab.

10. Click "Draw" in the tab "Visualize Count" to plot total and generation-specific cell count during time course (Fig. 6a). Experimental cell count is indicated by a red dot within the black line at each time point and a black line indicates the total cell count achieved by model fitting (*see* **Note 27**).

11. Comparison of multiple conditions can be achieved through the "Compare Phenotypes" tab (Fig. 6a) (*see* **Note 28**).

A

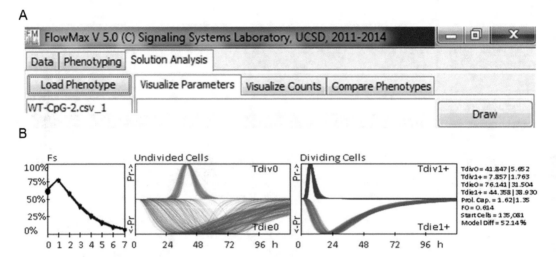

Fig. 6 Analyzing the cell biological parameter. (a) After the completion of the phenotyping go to the tab "Solution Analysis" and click load phenotype. Select the CSV file and click "Draw" (Draw is present at the top right corner) to visualize the cell biological parameters. (b) Cell biological parameter (FO, Tdiv0, Tdie0, Tdiv1 +, Tdie1+) obtained by FlowMax fitting. *X* axis refers to % of cells dividing and *Y* axis refers generation of cells. Fs: The bullet point in the *black curve* shows the % of cells dividing (Fs) in each generation. Undivided Cells: The distribution of the *green curve* shows the probability of time of division of generation "0" (Tdiv0) cells and distribution of the *red curve* shows the probability of time of die of generation "0" (Tdie0) cells. Dividing Cells: The distribution of the *blue curve* shows the probability of time of division of all the generation after "0" (Tdiv1+) cells and distribution of the *red curve* shows the probability of time of die of all the generation after "0" (Tdie1+) cells. The values of the cell biological parameters are shown in the *right side*. Though, it is better not to consider the Tdiv/Tdie value rather consider the distribution of Tdiv/Tdie in the curve respective curve

12. Cell biological parameters (F0, Tdiv0, Tdie0, Tdiv1+, Tdie1+) appear in the right side of the plot (Fig. 6b). These values are automatically saved as csv files in the working directory. Fs refers to the percentage of cells dividing. The black curve under Fs shows the percentage of cells dividing for each generation. F0 = 0.596 indicates ~59% of initial cells will divide. The green curve under undivided cells shows the probability density distribution of Tdiv0 within the B cell population. Tdiv0 = 42.462|4.727 indicates time of division of "generation 0" is 42.462 with standard deviation 4.727. Similarly, Tdie0, Tdiv1+, and Tdie1+ indicate the time to die of the initial generation, time of division of generations after the initial generation, and time to die of generations after the initial generation respectively. A curve is plotted for each model fit to indicate how well the distributions are constrained by the data. If the distributions are poorly constrained, represented by multiple non-overlapping curves, this may indicate that there is not enough information to accurately determine that biological process.

4   Notes

1. Prepare fresh media before (maximum 1 day earlier) isolation of B cells.

2. Keep the cold PBS at 37 °C water bath for 1 h to achieve (RT). Check the temperature of the media by touching the container using the palm of your hand. Only use it if the temperature is close to RT.

3. Keep the cold media for 2 h at RT. Check the temperature of the media by touching the container using the palm of your hand. Only use it if the temperature is close to RT.

4. Prepare freshly and store it at −80 °C in a 5 μL aliquot.

5. Usually the color of the spleen is reddish black. In some conditions it is common to see that a portion of spleen is black in color. The black part of the spleen is primarily dead cells. Do not take the black part of the spleen for cell isolation to increase the proportion of viable B cells.

6. Use 5 mL/mice, if spleen size is normal, i.e., total splenocytes $<250 \times 10^6$.

7. Do not use cold RBC lysis buffer. Keep the RBC lysis buffer for a maximum of 30 min in 37 °C water bath to reach RT. Check the temperature of the RBC lysis buffer by touching the container using the palm of your hand. Only use it if the temperature is close to RT.

8. If the pellet color is still red then repeat **steps 5** and **6** of Subheading 3.1.

9. Make a homogeneous suspension of the microbeads by vortexing the bottle.

10. Scale up the volume of the MACS buffer depending on the number of splenocytes.

11. CellTrace™ Violet/CellTrace™ Yellow/CFSE can be used as an alternate to CellTrace™ Far Red depending on the configuration of flow cytometry to be used. We observe that Cell-Trace™ conjugated dye shows more distinct peaks of different generations than CFSE. Also, the fluorescence decay rate of CFSE label is faster than CTR. CFSE labeled B cells lose fluorescence intensity rapidly in the first 48 h and more slowly thereafter.

12. The freezing point of DMSO 18.5 °C. If the RT is below 20 °C thaw it in >20 °C in a water bath.

13. Adjust the number of cells and CTR solution depending on the experimental need; e.g., if the experiment needs 10×10^6 then add 1 mL of working concentration of cells to 1 mL CTR solution.

14. Propidium Iodide (PI) can also be used as a dead cell marker.

15. It is important to count cells in the same volume for each time point. If 100 μL contains too many or too few cells adjust the count volume to get a significant number of cells.

16. FlowMax can work with FCS3.0 and FCS2.0 input files; however, it uses Accuri C6 metadata tags on the total volume collected to determine cell count automatically. For flow cytometers without accurate volume uptake measurement, we recommend manually measuring the volume of sample acquired and then manually normalizing the count from within FlowMax.

17. The 14 h time point is critical to get the correct value of F0. Lymphocytes isolation from tissue, such as spleen, bone marrow, etc., requires a mechanical tissue dissociation process that leads to cell death termed mechanical cell death [8]. Mechanical cell death occurs before 12 h [3]. To exclude mechanical cell death from the downstream analysis, the 14 h time point is critical. To get consistency of the data (F0, Tdiv0, and Tdie0) the first time point (14 h) has to be consistent in each replicate experiment.

18. The time points near the first division when cells have started dividing (usually between 24 and 48 h) and when most cells have died (usually 120–140 h after stimulation) are important for constraining the parameters during fitting. The time to first division and time when most cells are dead varies depending on the stimulus. Optimization of time points is required to achieve good FlowMax fitting.

19. Type "java" in the command prompt and a description of expected inputs and outputs should be displayed. If the "java" is not recognized, you may need to add its location to the $PATH variable of your operating system of use the full path to java (e.g., "/usr/bin/java").

20. FlowMax needs to load all FCS files into memory. This may require allocating additional memory for FlowMax. If analysis fails with larger datasets run FlowMax from command prompt using command "java –jar –Xmx1024m FlowMax.jar." Also, make sure that FlowMax is not running from a restricted directory such as "Program Files" in Windows.

21. To minimize fluorescence "spillover" from one channel into the next, a percentage of the fluorescence signal from one channel can be subtracted from another channel. First double-click on the sample you would like to compensate on the left. Then change the X- and Y-axis variables to the two fluorescence channels in which "spillover" is suspected. Then click on the multi-colored C next to each FCS sample in the left panel (Fig. 4a), adding the appropriate rules and clicking

"Compensate" (Fig. 4b). Repeat until you see a separation in the signals.

22. CTR label B cells gradually lose fluorescence of CTR during an experimental time course, probably due to catabolism of the CTR labeled protein. The result of which is a slight decrease in the intensity of generation "0" (undivided cell) peak as the time point increases. Accurately defining the generation "0" peak is very important to reliably estimate cell biological parameters. The best control, to correctly define generation "0," is unstimulated B cells. The rate of loss of fluorescence decay of unstimulated B cells is similar to stimulated B cells [5, 9].

23. Fitting proceeds in two steps, first fluorescence parameters are fitted to each histogram to account for experimental variability and loss of dye labeling over time. Next, the population parameters are repeatedly determined, using a stochastic optimization approach. An initial quick fit is recommended to see if reasonable solutions can be found using default parameter ranges.

24. The default parameter ranges may or may not be optimal for all conditions, and may need to be fine-tuned. To do this, check the box "Advanced Options" and tune the default parameter ranges. Hovering over each box gives a brief description. Constraining or relaxing some parameter ranges can help guide the optimization algorithm, however, over-zealous constraining will bias the result, and is not recommended without biological justification. We recommend listing the parameter ranges used and justifying changes when reporting results. For a full description of the fcyton model parameters, please *see* ref. 6.

25. Running 500–1000 solutions typically produces at least one good solution, but is time consuming. While running a few solutions should be enough for testing if parameter ranges are acceptable, we recommend running at least 500 solutions prior to interpretation of results.

26. When overlaying multiple phenotypes in the graph, you can change the color of each curve manually by clicking on the appropriate Overlay button and selecting the color of your choice. Uncheck the plot clusters box to plot the individual clustered solutions (e.g., the 500 solutions obtained), instead of sampling from the maximum-likelihood cluster ranges for parameters (default behavior).

27. The time axis, cell count axis, number of generations plotted, number of samples plotted, and the size of the graph, and whether counts are sampled or taken directly from the best-fit solutions in the cluster can be changed using the right panel. Select the fcyton model to ensure the right counts are plotted.

28. To determine which population parameters are sufficient for describing the difference between two phenotypes, FlowMax can be used to generate "chimeric" phenotypes containing parameters from both phenotypes. Check the parameters you would like to copy from the second phenotype, and hit the "Generate intermediate phenotype" button. FlowMax will create a new phenotype containing the parameters from the first phenotype with the selected parameters taken from the second phenotype. You can now plot the total cell counts to visually compare if the changes are sufficient for describing the population behavior.

Acknowledgment

This work was supported by NIH grant R01 AI132731 (A.H.) and from NIH-NCI CCSG: P30 014195, and the Helmsley Trust (M.N.S.).

References

1. de Fries R, Mitsuhashi M (1995) Quantification of mitogen induced human lymphocyte proliferation: comparison of alamarBlue assay to 3H-thymidine incorporation assay. J Clin Lab Anal 9:89–95

2. Denizot F, Lang R (1986) Rapid colorimetric assay for cell growth and survival. Modifications to the tetrazolium dye procedure giving improved sensitivity and reliability. J Immunol Methods 89:271–277

3. Hawkins ED, Hommel M, Turner ML, Battye FL, Markham JF et al (2007) Measuring lymphocyte proliferation, survival and differentiation using CFSE time-series data. Nat Protoc 2:2057–2067

4. Quah BJ, Warren HS, Parish CR (2007) Monitoring lymphocyte proliferation in vitro and in vivo with the intracellular fluorescent dye carboxyfluorescein diacetate succinimidyl ester. Nat Protoc 2:2049–2056

5. Hasbold J, Gett AV, Rush JS, Deenick E, Avery D et al (1999) Quantitative analysis of lymphocyte differentiation and proliferation in vitro using carboxyfluorescein diacetate succinimidyl ester. Immunol Cell Biol 77:516–522

6. Shokhirev MN, Hoffmann A (2013) FlowMax: a computational tool for maximum likelihood Deconvolution of CFSE time courses. PLoS One 8:e67620

7. Teodorovic LS, Riccardi C, Torres RM, Pelanda R (2012) Murine B cell development and antibody responses to model antigens are not impaired in the absence of the TNF receptor GITR. PLoS One 7:e31632

8. Klein AB, Witonsky SG, Ahmed SA, Holladay SD, Gogal RM Jr et al (2006) Impact of different cell isolation techniques on lymphocyte viability and function. J Immunoassay Immunochem 27:61–76

9. Lyons AB, Hasbold J, Hodgkin PD (2001) Flow cytometric analysis of cell division history using dilution of carboxyfluorescein diacetate succinimidyl ester, a stably integrated fluorescent probe. Methods Cell Biol 63:375–398

Chapter 7

Flow Cytometry Analysis of mTOR Signaling in Antigen-Specific B Cells

Qizhao Huang, Haoqiang Wang, Lifan Xu, Jianjun Hu, Pengcheng Wang, Yiding Li, and Lilin Ye

Abstract

B lymphocytes and their differentiated daughter cells are charged with responding to invading pathogens and producing protective antibodies against these pathogens. The physiology of B cells is intimately connected with the function of the B cell antigen receptor (BCR). Upon activation of BCR, transmembrane signals are generated, and several downstream pathways are activated, which provide a primary directive for the cell's subsequent response. mTOR is a serine/threonine kinase that controls cell proliferation and metabolism in response to a diverse range of extracellular stimuli. The activation of mTOR signaling downstream of PI3K/Akt activity by B cell receptor (BCR) engagement has been generally assumed to be essential for B cell responses. This chapter seeks to present two protocols to evaluate mTOR activity in B cells bearing BCR specific to 4-hydroxy-3-nitrophenylacetyl (NP)-hapten.

Key words Antigen specific B cell, B cell receptor, mTOR, NP-hapten, Ex-vivo, In vitro, Flow cytometry

1 Introduction

B cells are central components of adaptive immunity and respond to pathogens by proliferating, differentiating, and producing antibodies. Upon encountering an antigen, naive B cells become activated and migrate from the B cell follicle to the T–B border, whereby these cells will intensively interact with cognate CD4 T cells [1]. Subsequently, T cell-assisted B cells will either differentiate into short-lived extrafollicular plasma cells or migrate into B cell follicles to initiate a germinal center (GC) reaction [2, 3]. In the GC, activated B cells undergo rapid proliferation and somatic hyper-mutation/affinity maturation, which establishes a high-affinity, long-lived memory pool that contains memory B cells within lymphoid tissues and plasma cells in the bone marrow [4]. Therefore, B cell responses involve a series of cellular events, including migration, growth/proliferation, and differentiation.

Chaohong Liu (ed.), *B Cell Receptor Signaling: Methods and Protocols*, Methods in Molecular Biology, vol. 1707,
https://doi.org/10.1007/978-1-4939-7474-0_7, © Springer Science+Business Media, LLC 2018

The development and function of B cells are primarily governed by signaling via BCR [5]. Upon engagement of BCR, several downstream pathways are activated, including PI3K/AKT, NF-κB, and ERK/MAPK [6]. PI3K activation promotes Ca^{2+} mobilization and activation of NF-κB-dependent transcription, events that are all essential for B cell proliferation. PI3K also initiates a distinct signaling pathway involving Akt and the mammalian target of rapamycin (mTOR) [7]. mTOR is a serine/threonine kinase that controls cell proliferation and metabolism in response to a diverse range of extracellular stimuli such as the availability of nutrients, growth factors, and stress [8, 9]. Since mTOR integrates various signals to dictate the fate of T cell differentiation [10–13], it is of great interest to investigate whether mTOR signaling also influences the differentiation outcomes of activated B cells at the T–B border or within the germinal center. Recent studies have examined the role of mTOR in B cells and found that BCR-dependent mTOR signaling is required for the production of high-affinity antibodies [14–16]. Notably, mTOR signaling pathways regulate BCR-dependent B cell responses but are dispensable for LPS-induced B cell responses [17].

mTOR can be assembled to form two structurally and functionally distinct complexes, mTORC1 and mTORC2. mTORC1 has an array of substrates, among which the most well studied are the ribosomal S6 kinases (S6K1 and S6K2) and the eIF4E-binding proteins (4EBPs) [8]. S6Ks promote protein and lipid synthesis, both of which are primarily dependent on mTORC1-mediated phosphorylation [18]. The phosphorylation of 4EBPs by mTORC1 blocks its ability to suppress eIF4E function in cap-dependent translation [19]. Thus, the strength of mTORC1 activity can be evaluated based on the phosphorylation levels of S6 and 4EBP-1. On the other hand, mTORC2 activity can be measured by analyzing mTORC2-dependent phosphorylation of the hydrophobic motif of various AGC family kinases—most notably the S473 site on Akt.

Following immunization with an NP-conjugated protein, NP-specific B cells become activated, over 90% of which bear the V186.2 heavy chain and are Igλ⁺ [20]. This model antigen was of great value in determining antigen-specific B cell responses, including BCR-affinity maturation driven by somatic hypermutations [21]. To more conveniently study NP-specific B cell response, a B1–8^{high} mouse strain was created by the targeted insertion of a V_H 186.2 J_H 2 gene into the H chain (HC) locus of 129/Ola ES cells [22]. B cells that coexpress this B1–8 HC transgene with an endogenous λ1-LC express an allotypic variant of the HC that confers the specificity to the hapten NP.

In this chapter, we describe two detailed protocols to evaluate mTOR activity in antigen-specific B cells ex vivo and in vitro using flow cytometry. By combining the adoptive transfer of B1–8^{high}

cells and NP-OVA immunization, we can analyze mTOR activity in antigen-specific GC B cells ex vivo. In addition, we provided a protocol analyzing mTORC1 signaling in B1–8^{high} cells cultured in vitro. These protocols will facilitate the analysis of mTOR signaling in antigen-specific B cells under various conditions.

2 Materials

2.1 Mice

1. B1–8^{high} donor mice [23] (CD45.1, Jackson No: 007594), 6–8 weeks old.

2. Recipient mice: C57/B6 mice (CD45.2), 6–8 weeks old.

All the mouse experiments were performed in accordance with the guidelines of the Institutional Animal Care and Use Committees of the Third Military Medical University.

2.2 Immunization

1. 1.1 mg/mL NP(16)-OVA (4-Hydroxy-3-nitrophenylacetyl-conjugated ovalbumin, Biosearch Technologies), diluted in PBS and stored at −20 °C.

2. Complete Freund's adjuvant (Sigma); store at 4 °C.

3. Phosphate-buffered saline (PBS); store at room temperature.

4. 3-way stopcock.

5. Syringe and needle for subcutaneous injections.

2.3 Spleen/Lymph Node Collection

1. Cell culture medium R10: RPMI 1640 supplemented with 10% fetal calf serum (FCS, Gibco), 10,000 U/mL penicillin, and 10,000 μg/mL streptomycin. Store at 4 °C.

2. PBS.

3. Red blood cell (RBC) lysis buffer (RBL): 155 mM NH$_4$Cl, 10 mM KHCO$_3$, and 0.1 mM EDTA; pH 7.3. Store at room temperature.

4. 15 and 50 mL sterile conical tubes for cell processing.

5. Syringes for processing the spleen.

6. 70 μm nylon cell strainers.

7. Glass slides with frosted ends.

8. φ6 cm cell culture dishes.

9. Eppendorf centrifuge 5810R and 5424R.

2.4 B Cell Enrichment

1. MACS buffer: 2% FBS, 1% BSA, 0.25 M EDTA in PBS. Store at 4 °C.

2. Beaver magnetic beads. Store at 4 °C.

3. Magnet.

4. 1.5 mL EP tube and 15 mL conical tube

5. Biotinylated antibodies:

Marker	Fluorophore	Clone	Dilution
Fc blockade (CD16/CD32)		2.4G2	1:200
CD4	Biotin	RM4-5	1:300
CD8	Biotin	53-6.7	1:300
CD11C	Biotin	N418	1:800
Ter119	Biotin	TER-119	1:800
NK1.1	Biotin	PK136	1:800
Ig κ chain	Biotin	187.1	1:400
Ly6G	Biotin	1A8	1:800

2.5 Flow Cytometry

1. FACS buffer: 2% FBS in PBS. Store at 4 °C.

2. BD Phosflow lyse/fix buffer: Dilute 5× stock solution with water; prepare fresh.

3. BD Phosflow Perm/wash buffer I: Dilute 10× stock solution with water; prepare fresh.

4. Conjugated antibodies for flow-cytometry:

Marker	Clone	Dilution
Fc blockade (CD16/CD32)	2.4G2	1:200
B220	RA3-6B2	1:200
CD19	6D5	1:100
NP(31)		1:100
GL-7	GL-7	1:100
IgG1	A85-1	1:100
Igλ	R26-46	1:100
CD45.1	A20	1:200
CD71	R17217	1:100
CD98	RL388	1:100
p-S6 (S235/236)	D57.2.2E	1:100
p-4E-BP1 (T36/37)	236B4	1:100
p-FoxO1/3a	9464S	1:100
p-Akt (S473)	4060S	1:100
Donkey anti-rabbit IgG	A21206	1:1000
Live/dead		1:200

3   Methods

3.1 Isolation of Spleen Lymphocytes from B1–8^{high} Mice

Note: These procedures must be performed in a hood to maintain sterility.

1. Put one sterile φ6 cm dish per mouse in an ice box and add 3 mL of RBL to the dish.

2. Euthanize the mice and harvest the spleens (*see* **Note 1**).

3. Generate a single-cell suspension by gently smashing the spleen (using the tip of a syringe) through a 70 μm cell strainer that has been previously wetted with R10 and then rinse with R10. Transfer the cell suspension into a new 15 mL conical tube and keep it on ice while processing the other spleens.

4. Centrifuge the suspension at 4 °C at 500 × g for 6 min.

5. Discard the supernatant and resuspend cells in a specific volume of MACS buffer. Remove any flocs and aliquot a small sample cells for pre-depletion via flow cytometry (*see* **Note 2**). Count the cell number.

6. Centrifuge the suspension again at 4 °C at 500 × g for 6 min. Discard the supernatant and keep the cells on ice.

3.2 B1–8^{high} Cell Enrichment (See Note 3)

1. Prepare a biotin-negative selection cocktail (300 μL per 5 × 10^7 cells). Dilute biotinylated antibodies in MACS buffer and mix thoroughly as follows: anti-CD4 1:300, anti-CD8 1:300, anti-NK1.1 1:800, anti-Ly6G 1:800, anti-CD11c 1:800, anti-Ter119 1:800, and anti-Ig κ chain 1:400.

2. Resuspend the cells with the antibody cocktail. Mix well and incubate on ice for 20–25 min with gentle vortexing every 6 min.

3. Bring the cell suspension to a total volume of 10 mL in MACS buffer. Wash the cells by gently pipetting 3–5 times. Then, centrifuge the suspension at 4 °C at 500 × g, for 6 min.

4. Discard the supernatant and resuspend the cells in MACS buffer (300 μL per 5 × 10^7 cells).

5. Thoroughly vortex the magnetic beads to ensure that they are uniformly suspended. Add the magnetic beads to the cell suspension (100 μL beads per 300 μL MACS) and mix well. Rotate this cell-beads mixture at 4 °C for 30 min.

6. Add 1 mL of MACS buffer to the cell suspension and gently mix by inverting several times. Place the tube into the magnet and set aside for 5 min at room temperature (*see* **Note 4**).

7. Aspirate the supernatant into a new 15 mL conical tube and wash the cells as described in **step 3**.

8. Count the cell number.

3.3 Check the Percentage of Antigen-Specific B1–8^{high} Cells by Flow Cytometry

1. Add 100 μL per well of the cell suspension to a 96-well round-bottom plate (1×10^6/well) followed by 100 μL of FACS buffer to each well (*see* **Note 5**) which ensures consistency among the different samples and concomitantly reduces the use of reagents.

2. Pellet the cells by centrifuging the plates at $750 \times$ g for 1 min. Decant the medium by quickly flicking the plate. Vortex the plate for several seconds to disperse the cell pellet.

3. Wash once with FACS buffer (200 μL/well). Centrifuge at $750 \times g$ for 1 min. Decant the medium and vortex as described in **step 2**.

4. Resuspend the cells in Fc-blockade (1:200 diluted in FACS buffer; 50 μL/well) and incubate for 30 min on ice.

5. Prepare the conjugated antibody cocktail as follows (diluted in FACS buffer): anti-Igλ, 1:100, anti-B220, 1:200; anti-CD19, 1:200; NP-PE, 1:100; anti-CD45.1, 1:200; live/dead, 1:200 and Fc-blockade, 1:200 (*see* **Note 6**). Vortex the mixture and centrifuge at $15,000 \times g$ for 3 min to pellet the particulates. Store the antibody cocktail on ice in the dark (*see* **Note 7**).

6. Add prepared Abs cocktail 50 μL to each well without washing. Mix well by pipetting up and down. Incubate for 30 min on ice in the dark.

7. Wash the cells twice with 200 μL of FACS buffer. Centrifuge at $750 \times g$ for 1 min. Decant the medium and vortex.

8. Resuspend the cells in 200 μL of 2% PFA and transfer the cell suspensions into their respective FACS tubes.

9. Run the samples on a flow cytometer and analyze the data using FlowJo software according to the gating strategy shown in Fig. 1.

10. Calculate the number of B1–8^{high} cells: the number of live/dead$^-$B220$^+$CD19$^+$Iglampda$^+$NP$^+$ cells divided by the number of live/dead$^-$ cells multiplied by the total cell number.

3.4 Analyzing mTOR in NP-Specific GC B Cells Ex Vivo

3.4.1 Adoptive Transfer of B1–8^{high} Cells

1. Adjust the concentration of B1–8^{high} cells to 4×10^5/mL with RPMI.

2. Transfer 0.5 mL of the cell suspension (2×10^5 cells/mouse, *see* **Note 8**) to sex-matched recipient mice (CD45.2) via caudal vein injection (*see* **Note 9**).

3.4.2 NP-OVA Immunization of Recipient Mice

(12–18 h after cell transfer)

1. Dilute NP-OVA to 1 mg/mL with PBS. Aspirate 2.5 mL into a 10 mL syringe and connect the syringe to one end of the 3-way stopcock to expunge any air.

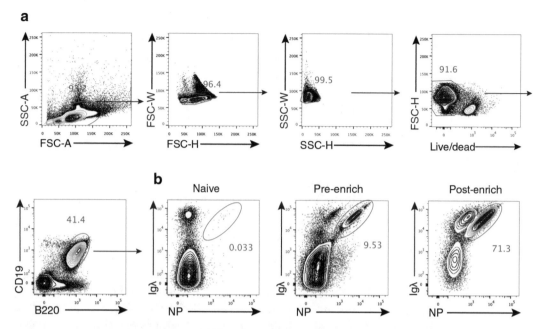

Fig. 1 Checking antigen-specific B1–8^{high} cell percentage post beads enrichment. The enrichment was performed as in protocol. (**a**) Flow cytometry gating strategy to identify NP$^+$Igλ$^+$ B cells (which stands for B1–8^{high} cells). (**b**) Flow cytometry analysis of NP$^+$Igλ$^+$ B1–8^{high} cells from pre-enrich and post-enrich groups. The *numbers* adjacent to the *outlined areas* indicate the proportion of each population

2. Thoroughly mix Complete Freund's Adjuvants and aspirate 2.5 mL into a 10 mL syringe. Add this to the other end of the 3-way stopcock.

3. Thoroughly emulsify these two solutions. The final concentration of NP-OVA is 0.5 mg/mL. After 20 min of mixing, the emulsion becomes difficult to move, and the color of emulsion becomes creamy (*see* **Note 10**).

4. Subcutaneously inject 100 μL (100 μg of protein/mouse) of the emulsion into each flank of the mouse. Make sure the emulsion is not injected intraperitoneally. The injection should form a visible bolus.

3.4.3 Cell Preparation from Lymph Nodes

Notice that the entire process must be done on ice (*see* **Note 11**).

1. Harvest inguinal lymph nodes from mice at day 8 post-immunization, and place lymph nodes in 6 cm cell culture dishes containing 3 mL RPMI supplemented with 2% FCS (R2).

2. Gently crush the dLNs using the frosted ends of glass slides. Make sure that all the dLNs are smashed.

3. Place a 70 μm strainer on the top of a 50 mL conical tube and filter the cell suspension.

4. Wash the cell strainer with 3 mL of R2.

5. Centrifuge at 4 °C at $500 \times g$ for 6 min.

6. Discard the supernatant and resuspend the cells in a specific volume of R2 (usually 1 mL). Carefully remove the floc.

7. Count the cell number. Usually, a single mouse provides 1×10^7 cells.

3.4.4 Surface Staining
(See **Note 12**)

1. Load 100 μL of cell suspension to a 96-well round-bottomed plate (1×10^6/well).

2. Wash once with FACS buffer (200 μL/well). Centrifuge at $750 \times g$ for 1 min. Decant the medium and vortex.

3. Resuspend the cells in Fc-blockade (1:200 diluted in FACS buffer; 50 μL/well) and incubate for 30 min on ice.

4. Make the antibody cocktail as follows (diluted in FACS buffer): anti-CD19, 1:200; anti-IgG1, 1:200; NP-PE, 1:100; anti-CD45.1, 1:200; anti-CD71, 1:100, anti-CD98, 1:100, GL-7, 1:100 and live/dead, 1:200 (CD71/CD98 can be allocated to another new panel for only 8 channels available in flow cytometer BD FACSCanto II). Vortex the mixture and centrifuge at $15,000 \times g$ for 3 min to pellet the particulates. Store the Abs cocktail on ice in the dark.

5. Stain the cells with the prepared Abs cocktail (50 μL/well) without washing. Mix well by pipetting up and down. Incubate for 30 min on ice in the dark.

6. Wash the cells twice with 200 μL of FACS buffer. Centrifuge at $750 \times g$ for 1 min. Decant the medium and vortex.

3.4.5 mTOR Signaling Staining

1. Pellet the cells by centrifuging at $750 \times g$ for 1 min. Decant the medium.

2. Immediately add 200 μL of pre-warmed BD lyse/fix buffer (5× stock, dilute with water) to each well and mix thoroughly by pipetting up and down several times.

3. Incubate at RT for 10–30 min.

4. Centrifuge at $750 \times g$ for 1 min and resuspend with Perm/Wash buffer I (10× stock, dilute with water).

5. Wash once with BD Perm/Wash buffer I.

6. Resuspend the pellet in BD Perm/Wash buffer I and incubate for 10 min at RT.

7. Centrifuge at $750 \times g$ for 1 min and decant the supernatant.

8. Add primary p-S6/p-4E-BP1/p-AKT/p-Foxo1a antibodies (all diluted 1:100 in BD Perm/Wash buffer I; 50 μL/well). Mix well by pipetting up and down (*see* **Note 13**).

9. Incubate at RT for 30 min in the dark.

10. Wash twice with BD Perm/Wash buffer I.

11. Add 50 µL/well of anti-Rabbit IgG antibody (1:1000 diluted in BD Perm/Wash buffer I) and anti-Mouse IgG1 (1:100 diluted in Perm/Wash buffer I). Mix well by pipetting up and down. Incubate at RT for 30 min in the dark.

12. Wash twice with BD Perm/Wash buffer I.

13. Wash once with FACS buffer.

14. Fix the cells in 2% PFA.

3.4.6 Flow Cytometer Analysis

1. Resuspend the cells in 200 µL of 2% PFA and transfer the cell suspensions into their respective FACS tubes.

2. Run the samples on a flow cytometer and analyze the data using FlowJo software (*see* **Note 14**).

3. The gating strategy is shown in Fig. 2 (*see* **Note 15**).

3.5 Analyzing mTOR Signaling Transduced by BCR Ligation In Vitro

1. Load 100 µL of B1–8^{high} cell suspension into a 96-well round-bottomed plate (1×10^6/well).

2. Wash once with FACS buffer (200 µL/well). Centrifuge at $750 \times g$ for 1 min. Decant the medium and vortex.

3.5.1 Surface Staining

3. Resuspend the cells in Fc-blockade (1:200 diluted in FACS buffer; 50 µL/well) and incubate for 30 min on ice.

4. Make the antibody cocktail as follows (diluted in FACS buffer): anti-B220, 1:200; anti-CD19, 1:200; anti-CD45.1, 1:200; and live/dead, 1:200. Vortex the mixture and centrifuge at $15,000 \times g$ for 3 min to pellet the particulates. Store the Abs cocktail on ice in the dark.

5. Stain the cells with the prepared Abs cocktail (50 µL/well) without washing. Mix well by pipetting up and down. Incubate for 30 min on ice in the dark.

6. Wash the cells twice with 200 µL of FACS buffer. Centrifuge at $750 \times g$ for 1 min. Decant the medium and vortex.

3.5.2 BCR Stimulation

1. Dilute the NP peptide to 100 µg/mL in the wells with a total volume of 200 µL of R10 per well. The control groups include NP peptide alone and non-stimulated cells. By comparing NP peptide stimulation with the non-stimulated group, the induction of mTOR signaling can be measured (*see* **Note 16**).

2. Place the cells in a 37 °C incubator containing 5% CO_2 for 0 h/ 2 h/8 h (*see* **Note 17**).

3.5.3 mTOR Signaling Staining

1. Pellet the cells by centrifuging at $750 \times g$ for 1 min. Decant the medium.

2. Immediately add 200 µL of pre-warmed BD lyse/fix buffer (5× stock, dilute with water) to each well and mix thoroughly by pipetting several times.

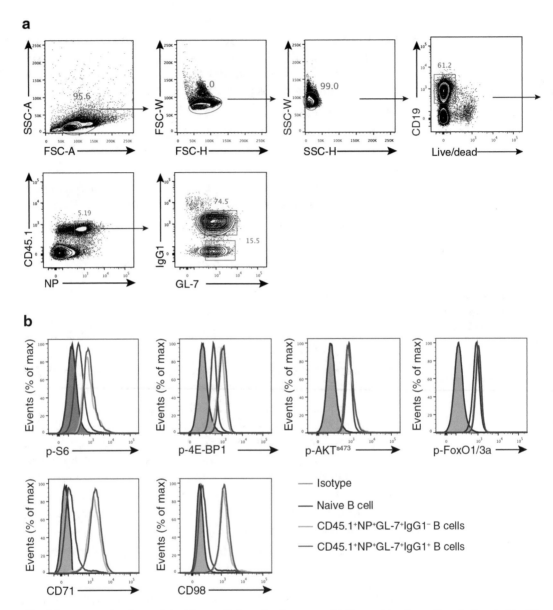

Fig. 2 Representative flow cytometric results from in-vivo assay. (**a**) Flow cytometry gating strategy to identify antigen-specific GL-7$^+$NP$^+$ B1–8^{high} cells (IgG1 and GL-7 expressing germinal center-like B cells have class switched). (**b**) Expression of p-S6, p-4E-BP1, p-AKT(s473),p-FoxO1/3a, CD71, and CD98 of B cells from different populations. *Red*, *green*, and *blue lines* represent the gating of IgG1$^+$GL-7$^+$NP$^+$, IgG1$^-$GL-7$^+$NP$^+$ and naive B cells respectively, and the *solid gray histograms* denote isotype control

3. Incubate at RT for 10–30 min.

4. Centrifuge at $750 \times g$ for 1 min, and resuspend with BD Perm/Wash buffer I (10× stock, dilute with water).

5. Wash with BD Perm/Wash buffer I once.

6. Resuspend with BD Perm/Wash buffer I and incubate for 10 min at RT.

7. Centrifuge at $750 \times g$ for 1 min and decant the supernatant.

8. Add primary p-S6/p-4E-BP1/p-AKT/p-Foxo1a antibodies (1:100 each, diluted in Perm/Wash buffer I; 50 μL/well). Mix well by pipetting up and down.

9. Incubate at RT for 30 min in the dark.

10. Wash twice with BD Perm/Wash buffer I.

11. Add anti-Rabbit IgG antibodies (1:1000 diluted in BD Perm/Wash buffer I; 50 μL/well). Mix well by pipetting up and down.

12. Wash twice with BD Perm/Wash buffer I.

13. Wash once with FACS buffer.

14. Fix the cells in 2% PFA.

3.5.4 Flow Cytometer Analysis

1. Resuspend the cells in 200 μL of 2% PFA and transfer the cell suspensions into their respective FACS tubes.

2. Run the samples on a flow cytometer and analyze the data using FlowJo software according to the gating strategy shown in Fig. 3.

4 Notes

1. The yield of B1–8^{high} cells greatly varies depending on the age of the mice, variation in preparing the emulsion, cell loss during preparation, and other factors. A typical yield is approximately 2×10^6 cells per mouse. Adjusting the number of mice may be necessary depending on the number of experimental parameters to be tested.

2. Save an aliquot of cells for pre-depletion as a control to compare the post-depleted cells to evaluate the efficiency of enrichment by comparing the frequency of B1–8^{high} cells between pre- and post-enrichment. Additionally, it is good practice to verify the correct genotype/phenotype of the cells prior to injection. This can be accomplished by quickly staining this small aliquot of cells. Stains containing congenic markers as well as the transgenic BCR should be conducted to verify the correct congenic markers and the presence of BCR-tg before injecting the cells into recipient mice.

3. In this chapter, we will describe the protocol from Beaver beads. Kits from other companies will also be valid but need to be used according to the manufacturer's instructions and optimized for a specific experimental design.

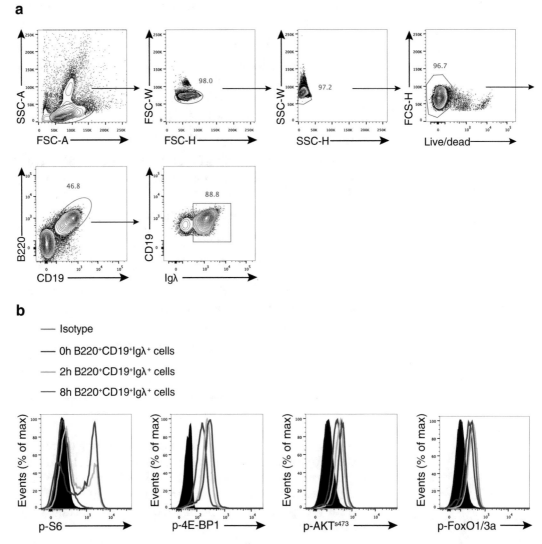

Fig. 3 Representative flow cytometric results from in-vitro assay. (**a**) Flow cytometry gating strategy to identify antigen-specific CD19⁺B220⁺Igλ⁺ B cells. (**b**) Expression of p-S6, p-4E-BP1, p-AKT(s473), and p-FoxO1/3a of B cells from different time points. Phosphorylated signaling proteins were stained after surfaces marker staining and then stimulated with NP-OVA or NP-PE in a 37 °C incubator for different hours. *Red, green,* and *blue lines* represent the gating of B220⁺CD19⁺Igλ⁺ B cells after 8 h, 2 h, and no stimulation respectively, and the *solid gray histograms* denote isotype control

4. To increase the efficiency of the enrichment, it is advisable to dilute the cell suspension by adding a specific volume of MACS buffer and aliquoting the cells into several tubes.

5. We found it convenient to perform the following steps by using round-bottom 96-well plates combined with multiple-channel pipette.

6. Prepare the antibody cocktail for all the samples together in an excess volume because pipetting errors may lead to an insufficient volume for the last samples.

7. High-speed centrifugation of diluted antibody mixes can remove fluorophore aggregates, which may appear as ultra-bright false positives when performing flow cytometry. Dilute the antibodies at the required dilution factor as determined by antibody titration, then centrifuge the mixture in a benchtop microfuge at maximum speed (up to $15,000 \times g$) for 3 min at 4 °C. After centrifuging the Abs cocktail, you can see colored particulates at the bottom of the tube. When adding the Abs cocktail to the cells, do not disturb the particulates pelleted in the bottom of the tube.

8. The number of cells prepared for transfer depends on the experimental parameters and time course. The number and frequency of NP-specific GC B cells will progressively increase from day 9 to day 28. A total of 2×10^5 cells/mouse are enough for a short-term experiment; for long-term adoptive transfer, the cell number transferred should be adjusted according to the specific experiments.

9. Adoptively transferred B1–8^{high} cells can be easily identified within recipient mice if they express distinct congenic or fluorescent markers. For example, different isoforms of the pan-lymphocyte marker CD45 (Ly5) are widely used. Other popular congenic markers include different isoforms of CD90 (Thy1). Crossing B1–8^{high} strains onto a CD45.1 or CD90.1 (Thy1.1) background provides the advantage of using normal C57BL/6 mice, which are CD45.2+ and CD90.2+, as host animals. Sex-matched recipients are required to avoid rejection issues.

10. Mix thoroughly, but cool to 4 °C if the syringes become warm, as excessive heat from friction will degrade the antigen. Check the emulsion by adding a small amount to PBS and make sure it does not partially dissolve. The emulsion should have the consistency of toothpaste.

11. Because phosphorylation is very unstable in a warm environment, it is advisable to precool the centrifuge and keep all the regents and cells on ice. Phosphatase inhibitors are widely used in protein extraction and should reduce dephosphorylation. In our experiment, however, we found that the presence or absence of a phosphatase inhibitor makes no difference.

12. Dead cells may exhibit significant autofluorescence and may nonspecifically bind antibodies. Thus, it is good practice to exclude dead cells during the analysis by incorporating a viability dye during the staining procedure. Here, we used a live/dead antibody to exclude dead cells by adding this antibody

with the other surface antibodies, and the live/dead negative cells are defined as live cells.

13. In this protocol, we first stain phosphorylated proteins with unconjugated primary antibodies of rabbit origin, followed by using fluorescein-labeled anti-rabbit secondary antibodies. Pay attention that two or more rabbits derived primary antibodies should not be included in the same antibody cocktails.

14. Before running the samples on the flow cytometer, be sure that you have samples of unstained cells and cells labeled with each individual mAb to allow for setup of the BD cell analyzer regarding setting the compensations.

15. It is good practice to have non-immunized control animals injected with either alum or vehicle only for each experimental group. This allows the precise gating of NP+ B cells.

16. For B1–8^{high} cell stimulation, NP-PE can also be used. Stimulating BCR with NP-OVA first may result in defective NP-PE and Igλ staining. Here, we stained for the surface markers, including Igλ, prior to stimulating with NP-OVA (or NP-PE) and immediately fixed the cells after stimulation.

17. One thing worth noting is that after a long period of BCR stimulation in a 37 °C incubator, the surface staining may be lost.

Acknowledgments

The work was supported by National Basic Research Program of China (973 program, 2013CB531500, to L.Y.), The National Key Research and Development Program of China (2016YFA0502202 to L.Y.).

References

1. Pereira JP, Kelly LM, Cyster JG (2010) Finding the right niche: B-cell migration in the early phases of T-dependent antibody responses. Int Immunol 22(6):413–419. https://doi.org/10.1093/intimm/dxq047

2. Mesin L, Ersching J, Victora GD (2016) Germinal center B cell dynamics. Immunity 45 (3):471–482. https://doi.org/10.1016/j.immuni.2016.09.001

3. Basso K, Dalla-Favera R (2015) Germinal centres and B cell lymphomagenesis. Nat Rev Immunol 15(3):172–184. https://doi.org/10.1038/nri3814

4. Gourley TS, Wherry EJ, Masopust D, Ahmed R (2004) Generation and maintenance of immunological memory. Semin Immunol 16 (5):323–333. https://doi.org/10.1016/j.smim.2004.08.013

5. Packard TA, Cambier JC (2013) B lymphocyte antigen receptor signaling: initiation, amplification, and regulation. F1000Prime Rep 5:40. 10.12703/P5-40

6. Seda V, Mraz M (2015) B-cell receptor signalling and its crosstalk with other pathways in normal and malignant cells. Eur J Haematol 94(3): 193–205. https://doi.org/10.1111/ejh.12427

7. Limon JJ, Fruman DA (2012) Akt and mTOR in B cell activation and differentiation. Front Immunol 3:228. https://doi.org/10.3389/fimmu.2012.00228

8. Laplante M, Sabatini DM (2012) mTOR signaling in growth control and disease. Cell 149

(2):274–293. https://doi.org/10.1016/j.cell.
2012.03.017

9. Zoncu R, Efeyan A, Sabatini DM (2011)
mTOR: from growth signal integration to cancer, diabetes and ageing. Nat Rev Mol Cell Biol
12(1):21–35.   https://doi.org/10.1038/
nrm3025

10. Pollizzi KN, Patel CH, Sun IH, Oh MH,
Waickman AT, Wen J, Delgoffe GM, Powell
JD (2015) mTORC1 and mTORC2 selectively
regulate CD8(+) T cell differentiation. J Clin
Invest 125(5):2090–2108. https://doi.org/
10.1172/JCI77746

11. Yang K, Shrestha S, Zeng H, Karmaus PW,
Neale G, Vogel P, Guertin DA, Lamb RF, Chi
H (2013) T cell exit from quiescence and differentiation into Th2 cells depend on Raptor-
mTORC1-mediated metabolic reprogramming. Immunity 39(6):1043–1056. https://
doi.org/10.1016/J.Immuni.2013.1009.
1015. Epub 2013 Dec 1045

12. Kim JS, Sklarz T, Banks LB, Gohil M, Waickman AT, Skuli N, Krock BL, Luo CT, Hu W,
Pollizzi KN, Li MO, Rathmell JC, Birnbaum
MJ, Powell JD, Jordan MS, Koretzky GA
(2013) Natural and inducible TH17 cells are
regulated differently by Akt and mTOR pathways. Nat Immunol 14(6):611–618. https://
doi.org/10.1038/ni.2607

13. Delgoffe GM, Kole TP, Zheng Y, Zarek PE,
Matthews KL, Xiao B, Worley PF, Kozma SC,
Powell JD (2009) The mTOR kinase differentially regulates effector and regulatory T cell
lineage   commitment.   Immunity   30
(6):832–844.   https://doi.org/10.1016/j.
immuni.2009.04.014

14. Zhang S, Pruitt M, Tran D, Du Bois W,
Zhang K, Patel R, Hoover S, Simpson RM,
Simmons J, Gary J, Snapper CM, Casellas R,
Mock BA (2013) B cell-specific deficiencies in
mTOR limit humoral immune responses. J
Immunol 191(4):1692–1703. https://doi.
org/10.4049/jimmunol.1201767

15. Keating R, Hertz T, Wehenkel M, Harris TL,
Edwards BA, McClaren JL, Brown SA,
Surman S, Wilson ZS, Bradley P, Hurwitz J,
Chi H, Doherty PC, Thomas PG, McGargill
MA (2013) The kinase mTOR modulates the
antibody response to provide cross-protective
immunity to lethal infection with influenza

virus. Nat Immunol 14(12):1266–1276.
https://doi.org/10.1038/ni.2741

16. Jones DD, Gaudette BT, Wilmore JR,
Chernova I, Bortnick A, Weiss BM, Allman D
(2016) mTOR has distinct functions in generating versus sustaining humoral immunity. J
Clin Invest 126(11):4250–4261. https://doi.
org/10.1172/JCI86504. Epub 82016 Oct
86517

17. Ye L, Lee J, Xu L, Mohammed AU, Li W, Hale
JS, Tan WG, Wu T, Davis CW, Ahmed R, Araki
K (2017) mTOR promotes antiviral Humoral
immunity by differentially regulating CD4
helper T cell and B cell responses. J Virol 91
(4):e01653–01616.   https://doi.org/10.
1128/JVI.01653-16. Print 02017 Feb 01615

18. Magnuson B, Ekim B, Fingar DC (2012) Regulation and function of ribosomal protein S6
kinase (S6K) within mTOR signalling networks. Biochem J 441(1):1–21. https://doi.
org/10.1042/BJ20110892

19. Silvera D, Formenti SC, Schneider RJ (2010)
Translational control in cancer. Nat Rev Cancer
10(4):254–266.   https://doi.org/10.1038/
nrc2824

20. Bothwell AL, Paskind M, Reth M, Imanishi-
Kari T, Rajewsky K, Baltimore D (1981) Heavy
chain variable region contribution to the NPb
family of antibodies: somatic mutation evident
in a gamma 2a variable region. Cell 24
(3):625–637

21. Lalor PA, Nossal GJ, Sanderson RD,
McHeyzer-Williams MG (1992) Functional
and molecular characterization of single,
(4-hydroxy-3-nitrophenyl)acetyl (NP)-specific,
IgG1+ B cells from antibody-secreting and
memory B cell pathways in the C57BL/6
immune response to NP. Eur J Immunol 22
(11):3001–3011. https://doi.org/10.1002/
eji.1830221136

22. Sonoda E, Pewzner-Jung Y, Schwers S, Taki S,
Jung S, Eilat D, Rajewsky K (1997) B cell
development under the condition of allelic
inclusion. Immunity 6(3):225–233

23. Shih TA, Roederer M, Nussenzweig MC
(2002) Role of antigen receptor affinity in T
cell-independent antibody responses in vivo.
Nat Immunol 3(4):399–406. https://doi.
org/10.1038/ni776

Chapter 8

Ex Vivo Culture Assay to Measure Human Follicular Helper T (Tfh) Cell-Mediated Human B Cell Proliferation and Differentiation

Xin Gao, Lin Lin, and Di Yu

Abstract

Upon priming by antigens, B cells undergo activation, proliferation, and differentiation into antibody-secreting cells. During thymus-dependent (TD) antibody responses, the proliferation and differentiation of antigen-primed B cells essentially rely on the helper function from CD4$^+$ T cells. Follicular helper T (Tfh) cells constitute a specialized Th subset that localizes in close proximity to B cells and supports B cell proliferation and differentiation through co-stimulatory receptors and cytokines. Impaired Tfh-mediated B cell proliferation and differentiation were observed in patients with immunodeficiency, while overactivation of this process may lead to dysregulated immune responses seen in autoimmune disorders. Here, we describe an ex vivo coculture assay using circulating Tfh cells and B cells isolated from human blood. This method can be used to examine the function of patients' B cells for proliferation, differentiation, and antibody secretion, mediated by the physiological help from Tfh cells.

Key words Follicular helper T cell, B cell, Coculture, Proliferation, Differentiation

1 Introduction

B cells secrete antibodies that confer humoral immunity to provide protection against a great variety of pathogens. In thymus-independent (TI) responses, B cells are stimulated via signals from toll-like receptors (TLRs) by ligands such as CpG or from cross-linked B cell receptors (BCRs) by polysaccharides. In contrast, in thymus-dependent (TD) responses, B cells recognize protein antigens and also critically rely on helper function from CD4$^+$ T cells to receive co-stimulatory and cytokine signals for a successful TD response [1, 2]. TD responses are considered to be the main path to generate affinity-maturated and isotype-switched antibodies and B cell memory, which underlie the success of most vaccination to date.

Chaohong Liu (ed.), *B Cell Receptor Signaling: Methods and Protocols*, Methods in Molecular Biology, vol. 1707, https://doi.org/10.1007/978-1-4939-7474-0_8, © Springer Science+Business Media, LLC 2018

Follicular helper T (Tfh) cells represent a specialized CD4$^+$ helper T cell subset to provide helper to B cells. Tfh cells express the chemokine receptor CXCR5 to migrate to B cell follicles and support B cell proliferation, germinal center formation, affinity maturation, and differentiation into antibody-secreting cells and memory cells. Through cognate interactions, Tfh cells provide co-stimulatory and cytokine signals to B cells, which synergize with BCR signaling for B cell survival, proliferation, and differentiation [3]. The defective signals from co-stimulatory or cytokine receptors on B cells are associated with impaired B cell function and can cause primary immunodeficiency. For example, patients with mutations in CD40 or CD40 ligand (CD154) developed the hyper IgM syndrome, characterized by defective germinal center formation and very low levels of IgG, IgA, and IgE, with a normal or elevated IgM [4]. Mutations in IL-21 or IL-21 receptor in patients also caused reduced B cell proliferation, class-switching, and differentiation [5].

It is of great importance to measure B cell proliferation, differentiation, and antibody secretion in basic research and clinical application such as examining patients' samples. Several methods have been developed to stimulate B cells ex vivo for proliferation and differentiation, such as by anti-IgM and/or anti-CD40 stimulating antibodies or TLR agonists, which stimulate naive and memory B cells through different signaling pathways [6–8]. However, none of these methods could closely resemble the cognate help from Tfh cells to facilitate B cell activation and differentiation.

In this chapter, we will describe a protocol for an ex vivo assay to coculture circulating Tfh cells [9] with naive or memory B cells isolated from human blood, and measure Tfh cell-mediated B cell proliferation and differentiation. The advantage of this method is its resemblance to the physiological condition by which Tfh cells help B cells. We will demonstrate Tfh cells greatly enhance the proliferation, plasmablast differentiation, and antibody secretion of B cells. Notably, naive and memory B cells show different levels of dependence on Tfh cells in these processes.

2   Materials

2.1   PBMC Preparation from Human Peripheral Blood

1. 10 mL K2-EDTA tube for blood collection.

2. Ficoll-Paque™PLUS or other density gradient solutions for human PBMC isolation.

3. Sterilized phosphate-buffered saline (PBS).

4. 50 mL sterile conical tubes.

Table 1
The staining panel for sorting circulating Tfh cells, naive and memory B cells

Antigen/reagent	Clone
TCRα/β	IP26
IgD	IA6-2
CD27	M-T271
CXCR5	RF8B2
CD45RA	HI100
CD4	OKT4
CD19	SJ25C1

2.2 Cell Sorting and Tfh-B Cell Coculture

1. A full staining panel of antibodies (Table 1).
2. CellTrace™ Violet (CTV) cell proliferation kit.
3. Sorting buffer: 5% (v/v) deactivated fetal bovine serum (FBS) in sterilized PBS.
4. 70 μm nylon cell strainer.
5. 5 mL sterile FACS tubes.
6. 96-well round-bottom cell culture plates.
7. BD FACS Aria III with four lasers (Blue/488 nm; Red/633 nm; Violet/405 nm; YellGrn/561 nm) or equivalent equipment.
8. Complete RPMI: 10% deactivated FBS (v/v), 100 units/mL penicillin, 100 μg/mL streptomycin, 1 mM sodium pyruvate, 1% MEM nonessential amino acids (v/v), and 0.055 mM β-Mercaptoethanol in RPMI 1640 with L-glutamine and 25 mM HEPES.
9. Staphylococcal enterotoxin (SEB).

2.3 FACS Analysis to Determine B Cell Proliferation and Differentiation

1. A full staining panel of antibodies (Table 2).
2. FACS buffer: 5% deactivated FBS (v/v) and 0.1% sodium azide (w/v) in sterilized PBS.
3. 96-well V bottom assay plates.
4. 5 mL sterile FACS tubes.
5. BD LSR Fortessa with four lasers (Blue/488 nm; Red/633 nm; Violet/405 nm; YellGrn/561 nm) or equivalent equipment.
6. FlowJo analysis software.

Table 2
The staining panel for FACS analysis

Antigen/reagent	Clone
7AAD	–
CD27	M-T271
CD38	HB7
CD4	OKT4

2.4 ELISA to Quantify Antibody Secreted by B Cells in Coculture Supernatant

1. 96-well high protein binding ELISA plates.
2. MABTECH human IgM, IgG, and IgA ELISA development kits.
3. PBST: 0.05% (v/v) Tween 20 in PBS.
4. Incubation buffer: 0.1% (w/v) bovine serum albumin (BSA) in PBST.
5. *p*-Nitrophenyl-phosphate (pNPP).
6. Spectrometer.

3 Methods

3.1 PBMC Preparation from Human Peripheral Blood (See Note 1)

1. Collect human periphery blood in 10 mL K2-EDTA tubes (*see* **Note 2**).
2. Transfer the blood to a 50 mL conical tube, add the same volume of sterilized PBS and gently mix.
3. Add Ficoll to another 50 mL conical tube. The volume equals the original peripheral blood.
4. Carefully layer the diluted blood onto Ficoll by gently dripping the blood onto the surface of Ficoll.
5. Centrifuge at $400 \times g$ for 30 min at room temperature. The parameters for acceleration and deceleration are set to 2 and 0.
6. Carefully transfer the isolated PBMC (*see* **Note 3**) into a 15 mL conical tube.
7. Wash the isolated PBMC by sterilized PBS once.

3.2 Cell Sorting and Tfh-B Cell Coculture

1. Prepare an antibody mixture with sorting buffer according to the panel (Table 1).
2. Resuspend the isolated human PBMC in the antibody mixture and incubate at room temperature for 20 min. The volume of antibody mixture is adjusted for cell numbers. Up to 10 million cells can be stained per 100 μL antibody mixture.
3. Wash the PBMC once with 10 mL sorting buffer and resuspend in 1 mL sorting buffer.

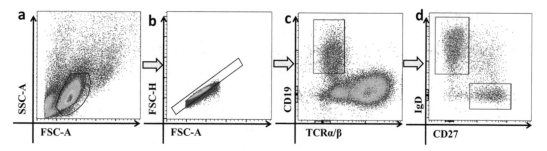

Fig. 1 The gating strategy for naive and memory B cells. (**a**) Forward scatter (FSC) and side scatter (SSC) were used to gate for lymphocytes. (**b**) FSC-A and FSC-H were used to gate for single cells. (**c**) B cells: CD19$^+$ TCRα/β$^-$. (**d**) Naive B cells: CD27$^-$ IgD$^+$; Memory B cells: CD27$^+$ IgD$^-$

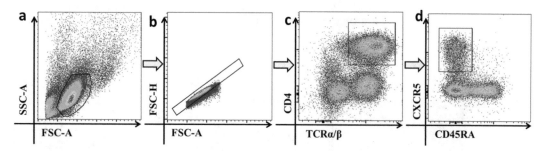

Fig. 2 The gating strategy for circulating Tfh cells. (**a**) FSC-A and SSC-A were used to gate for lymphocytes. (**b**) FSC-A and FSC-H were used to gate for single cells. (**c**) CD4$^+$ T cells: TCRα/β$^+$ CD4$^+$. (**d**) Circulating Tfh cells: CD45RA$^-$ CXCR5$^+$

4. Filter the PBMC through a 70 μm strainer to a 5 mL sterile FACS tube.

5. Load the FACS tube onto the BD Aria III and sort for circulating Tfh cells, naive and memory B cells. The gating strategies are shown in Figs. 1 and 2.

6. Collect circulating Tfh cells, naive and memory B cells and wash once with 10 mL sorting buffer (*see* **Note 4**).

7. Label sorted naive B cells and memory B cells by CTV according to the manufacturer's instruction (*see* **Note 5**).

8. Resuspend circulating Tfh cells, naive and memory B cells in 200 μL complete RPMI.

9. Count cell numbers and set up the Tfh-B cell coculture assay in a 96-well round-bottom cell culture plate. Add 50,000 circulating Tfh cells and 50,000 naive or memory B cells in each well. As controls, only naive or memory B cells will be added without circulating Tfh cells. Add SEB (100 ng/mL) to the coculture and adjust the volume to 200 μL. Prepare triplicates for every experimental condition.

10. Incubate the plate in a tissue culture incubator (5% CO_2 and 37 °C) for 6 days (*see* **Note 6**).

3.3 FACS Analysis to Determine Tfh Cell-Mediated B Cell Proliferation and Differentiation

1. Prepare an antibody mixture with FACS buffer according to the panel (Table 2).

2. Collect the culture supernatant for ELISA tests.

3. Transfer the cells from each well to a 96-well V-bottom assay plate and wash the cells once with 200 μL FACS buffer.

4. Resuspend the cells in 50 μL antibody mixture per well and incubate on ice for 25 min.

5. Wash the cells once with 200 μL FACS buffer and resuspend in 200 μL FACS buffer.

6. Transfer the cells into 5 mL FACS tubes and perform FACS analysis.

7. Analyze the samples according to the gating strategy shown in Fig. 3. The examples of the results are shown in Figs. 4 and 5.

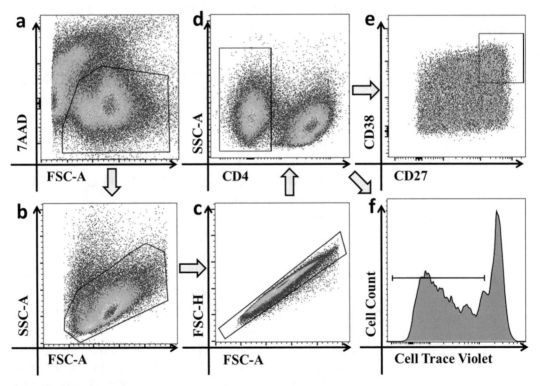

Fig. 3 The gating strategy for FACS analysis of cocultured cells. The example from a coculture of circulating Tfh cells and memory B cells. (**a**) FSC-A and 7AAD were used to gate for viable cells: 7AAD⁻. (**b**) FSC-A and SSC-A were used to gate for lymphocytes. (**c**) FSC-A and FSC-H were used to gate for single cells. (**d**) Exclusion of CD4⁺ T cells for analysis. (**e**) Plasmablasts: CD27^high CD38^high. (**f**) A histogram of CTV to quantify B cell proliferation

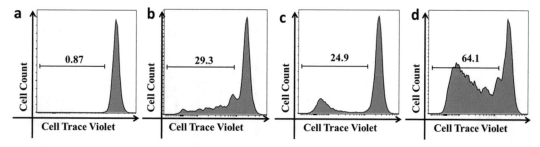

Fig. 4 Examples of Tfh cell-mediated proliferation of naive or memory B cells. Naive B cells or memory B cells were cocultured with circulating Tfh cells at 1:1 ratio for 6 days, with no Tfh cells as controls. FACS analysis showed CTV-label for B cell proliferation. (**a**) Naive B cells alone. (**b**) Naive B cells with Tfh cells. (**c**) Memory B cells alone. (**d**) Memory B cells with Tfh cells

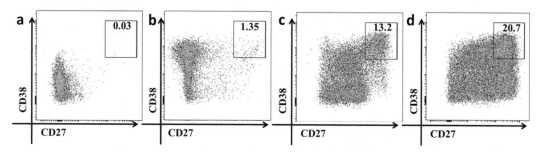

Fig. 5 Examples of Tfh cell-mediated plasmablast differentiation of naive B cells or memory B cells. Naive or memory B cells were cocultured with circulating Tfh cells at 1:1 ratio for 6 days, with no Tfh cells as controls. FACS analysis showed the differentiation of CD38^{high} CD27^{high} plasmablasts. (**a**) Naive B cells alone. (**b**) Naive B cells with Tfh cells. (**c**) Memory B cells alone. (**d**) Memory B cells with Tfh cells

3.4 ELISA to Quantify Antibodies Secreted by B Cells in Coculture Supernatant

1. Coat a 96-well high protein binding ELISA plate with each of anti-IgM, IgG, or IgA antibodies overnight at 4 °C.

2. Wash twice with 200 μL PBS.

3. Block plate by adding 200 μL of incubation buffer for 1 h at room temperature.

4. Wash twice with PBST.

5. Add 50 μL per well of coculture supernatants as well as IgM, IgG, or IgA standards and incubate for 2 h at room temperature (*see* **Note 7**).

6. Wash twice with 200 μL PBST.

7. Add 50 μL per well of anti-IgM, IgG, or IgA-alkaline phosphatase (×1000 dilutions) in incubation buffer and incubate for 1 h at room temperature.

8. Wash five times with 200 μL PBST.

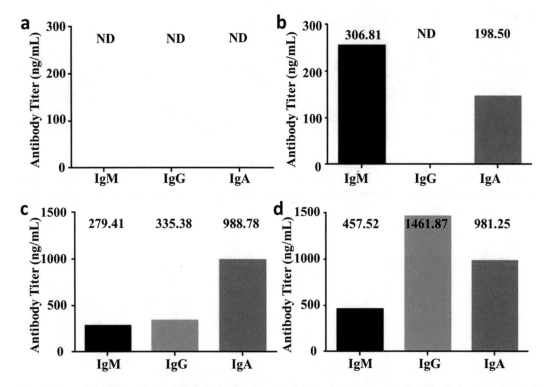

Fig. 6 Examples of Tfh cell-mediated antibody secretion from naive or memory B cells. Naive or memory B cells were cocultured with circulating Tfh cells at 1:1 ratio for 6 days, with no Tfh cells as controls. Supernatants were analyzed by ELISA. (**a**) Naive B cells alone. (**b**) Naive B cells with Tfh cells. (**c**) Memory B cells alone. (**d**) Memory B cells with Tfh cells. *ND* not detectable

9. Add 50 μL per well of pNPP.

10. Measure the absorbance for 405 nm using a spectrometer. The examples of results are shown in Fig. 6.

---

4   Notes

1. Procedures in Subheadings 3.1 and 3.2 should be conducted in biosafety hood to avoid contamination.

2. The volume of blood required for an experiment can vary. It depends on the yield of T cells and B cells and also the conditions of the experiments.

3. PBMC are in the semitransparent white layer between Ficoll and plasma.

4. We can usually yield more than 400,000 circulating Tfh cells, 400,000 naive and 200,000 memory B cells from 20 mL blood.

5. Homogenous staining for CTV on all the cells is critical to clearly define divided cells. When many samples are stained

simultaneously, we recommend carrying out the staining procedure on a 96-well V-bottom culture plate using a multichannel pipette.

6. A gentle pipetting of cocultured cells on day 3 can help to re-mix Tfh cells and B cells.

7. According to the manufacturer's protocol, measured samples should range from 0.1 to 500 ng/mL for accurate ELISA results. According to our experience, coculture supernatants need to dilute 2–5 times to fall into this range for ELISA tests.

Acknowledgment

This work was supported by the National Natural Science Foundation of China (NSFC) (81429003), Start-up Funding from Renji Hospital, Shanghai Jiaotong University School of Medicine, Australian National Health and Medical Research Council Fellowship (GNT1085509) and the Australian National University Future Scheme to D.Y.

References

1. Parker DC (1992) T cell-dependent B cell activation. Annu Rev Immunol 11(1):331

2. Lanzavecchia A, Sallusto F (2007) Toll-like receptors and innate immunity in B-cell activation and antibody responses. Curr Opin Immunol 19(3):268–274

3. Vinuesa CG, Linterman MA, Yu D, MacLennan IC (2016) Follicular helper T cells. Annu Rev Immunol 34:335–368

4. Lougaris V, Badolato R, Ferrari S, Plebani A (2005) Hyper immunoglobulin M syndrome due to CD40 deficiency: clinical, molecular, and immunological features. Immunol Rev 203 (1):48–66

5. Kotlarz D, Zietara N, Milner JD, Klein C (2014) Human IL-21 and IL-21R deficiencies: two novel entities of primary immunodeficiency. Curr Opin Pediatr 26(6):704–712

6. Cao Y, Gordic M, Kobold S, Lajmi N, Meyer S, Bartels K, Bokemeyer C (2010) An optimized assay for the enumeration of antigen-specific memory B cells in different compartments of the human body. J Immunol Methods 358 (1):56–65. https://doi.org/10.1016/j.jim.2010.03.009

7. Hanten JA, Vasilakos JP, Riter CL, Neys L, Lipson KE, Alkan SS, Birmachu W (2008) Comparison of human B cell activation by TLR7 and TLR9 agonists. BMC Immunol 9(1):39. https://doi.org/10.1186/1471-2172-9-39

8. Fecteau JF, Néron S (2003) CD40 stimulation of human peripheral B lymphocytes: distinct response from naive and memory cells. J Immunol 171(9):4621–4629

9. Wei Y, Feng J, Hou Z, Wang XM, Yu D (2015) Flow cytometric analysis of circulating follicular helper T (tfh) and follicular regulatory T (tfr) populations in human blood. Methods Mol Biol 1291:199–207

Chapter 9

B Cell Receptor Signaling and Compartmentalization by Confocal Microscopy

Anurag R. Mishra and Akanksha Chaturvedi

Abstract

Binding of antigen to the B cell receptor (BCR) triggers both BCR signaling and endocytosis simultaneously. BCR signaling pathways and their regulation have been studied extensively by both biochemical methods and flow cytometry, resulting in a comprehensive understanding of the temporal dynamics of the signaling enzymes and effector proteins. However, spatial regulation of these signaling pathways in subcellular pathways is relatively poorly understood. Here, we describe a method to study the spatiotemporal distribution of phosphorylated-kinases in antigen-activated B cells by confocal microscopy. This method can also be applied to other cell types where it is of interest to understand the spatial distribution of signaling enzymes and their effector proteins.

Key words B cells, B cell receptors, BCR signaling, Endocytosis, Confocal microscopy

1 Introduction

B cells are crucial player of the adaptive immune system, which recognize a plethora of antigens and respond by secreting highly specific antibodies against them [1]. B cell activation is initiated by the binding of antigens (Ags) to the clonally expressed B cell receptors (BCRs) on their surface. The BCR consists of a membrane-associated form of immunoglobulin (Ig) bound non-covalently with an Igα-Igβ heterodimer [2]. Naive B cells express both membrane IgM and IgD BCRs. Both Igα-Igβ contain immunoreceptor tyrosine activation motifs (ITAMs) in their cytoplasmic domains. Following Ag binding, ITAMs are phosphorylated by Src family protein tyrosine kinases (PTKs) such as Lyn leading to the phosphorylation of Syk and recruitment of the activated Syk to the BCR signalosome. This results in the initiation of downstream signaling cascades including activation of MAP family kinases, e.g., ERK, JNK, and p38 [3–5]. Ag-bound BCR is also endocytosed simultaneously and traffics through TfR+ early endosomes to LAMP1+ late endosomes and multivesicular bodies

Chaohong Liu (ed.), *B Cell Receptor Signaling: Methods and Protocols*, Methods in Molecular Biology, vol. 1707,
https://doi.org/10.1007/978-1-4939-7474-0_9, © Springer Science+Business Media, LLC 2018

(MVBs) where the Ag is processed and presented on MHC class II molecules for recognition by helper T cells [6, 7].

BCR signaling is a complex and dynamic process that involves sequential activation of a vast number of kinases, signaling adaptors, second messengers, and phosphatases [3–5]. The temporal regulation of the activation of many components of the BCR signaling pathways has been described in great detail by flow cytometry and biochemical methods. However, one of the caveats of these techniques is that they do not take into consideration where in the B cells each of these discrete signaling events takes place. In this method, we labeled the B cells with fluorophore-labeled specific antibodies against phosphorylated kinases, endosomal and lysosomal markers, TfR and LAMP1 respectively, and imaged them with confocal laser scanning microscopy to identify the subcellular location where these kinases are accumulated in phosphorylated form following BCR signaling. BCR upon Ag binding leads to sequential recruitment and phosphorylation of signaling kinases beginning at the plasma membrane and later as the BCR traffics to the early and late endosomal compartments [8]. Confocal laser scanning microscopy is a preferred method to visualize intracellular localization of proteins in both living and fixed cells. Confocal microscopy has been used widely to establish that the endo-lysosomal compartments are an essential site of signal transduction and to map receptor trafficking patterns for a number of receptors [9–16]. Receptor trafficking regulates signaling partly by compartmentalizing the scaffolding proteins and signaling components in distinct subcellular compartments, thereby controlling the access of receptor to signaling intermediates [17, 18]. This method described below provides a mean to study the spatial and temporal relationship between the endocytosis of the BCR and the assembly of active signaling complexes [8]. Although this method is optimized for the imaging of phosphorylated kinases in B cells, it is applicable to other signaling components and other cell types as well.

2 Materials

1. Purified B cells (*see* **Note 1**).

2. RPMI 1640 (Thermofisher Scientific, MA, USA). Supplement RPMI with 10 mM of MEM nonessential amino acids solution (Thermofisher Scientific, MA, USA), 50 mM β-mercaptoethanol, 10,000 units/mL of penicillin and 10,000 μg/mL of streptomycin solution (Pen Strep, 100 X) (Thermofisher Scientific, MA, USA).

3. Fetal bovine serum, heat inactivated (Thermofisher Scientific, MA, USA).

4. Bovine serum albumin.

5. 15 mL Falcon tubes.

6. BioCoat Poly-D-Lysine coated 8 well Culture Slide (Corning®️ BioCoat™️) (*see* **Note 2**).

7. 24 × 50 mm #1.5 coverglasses.

8. Phosphate-buffered saline (PBS), pH 7.4. Prepare 10 X stock with 1.55 M NaCl, 16.9 mM KH_2PO_4, and 26.5 mM Na_2HPO_4, pH 7.4 (adjusted with HCl), dilute with ultrapure deionized water, and filter with a disposable bottle-top vacuum 0.22 μm filter.

9. Triton X-100.

10. 16% paraformaldehyde solution.

11. Deionized ultrapure water.

12. Permeabilization solution—1 X PBS, 0.2% Triton X-100.

13. Blocking buffer—1 X PBS, 0.2% Triton X-100 2% BSA, 50 μg/mL normal goat serum.

14. Alexa 647 labeled goat F(ab)$_2$ anti-mouse IgM (Jackson Immuno Research Lab, West Grove, PA). Prepare 1 mg/mL stock solution, in 50% glycerol, make aliquots and store at −20 °C. Stable for at least 1 year in this condition.

15. Unlabeled goat F(ab)$_2$ anti-mouse IgM (Jackson Immuno Research Lab, West Grove, PA). Prepare 1 mg/mL stock solution and store at 4 °C.

16. Normal goat serum.

17. Primary antibodies (*see* Table 1).

Table 1
List of antibodies

List of antibodies	Supplier	Dilution
Rat anti-mouse TfR (C2F2)	BD Biosciences	1:100
Rat anti-mouse LAMP-1 (1D4B)	Santa Cruz	1:100
Alexa 488-conjugated goat F(ab)$_2$ Ab to rabbit Ig	Thermo Fisher Scientific	1:2000
Alexa 488-conjugated goat F(ab)$_2$ Ab to rat Ig	Thermo Fisher Scientific	1:2000
Alexa 647-conjugated goat F(ab)$_2$ Ab to rabbit Ig	Thermo Fisher Scientific	1:2000
Rabbit Abs to phospho-Lyn (Cat # 2731)	Cell signaling	1:200
Rabbit Abs to phospho-Syk (Cat # 2701)	Cell signaling	1:200
Rabbit Abs to phospho-Jnk (Cat # 9251)	Cell signaling	1:200
Rabbit Abs to phospho-ERK (Cat # AF1018)	R&D Systems Inc.	1:100
Normal goat serum	Jackson ImmunoResearch	50 μg/mL
Alexa 647–F(ab)$_2$ goat antibody to mouse IgM	Jackson ImmunoResearch	10 μg/mL
F(ab)$_2$ goat antibody to mouse IgM, unconjugated	Jackson ImmunoResearch	10 μg/mL

18. Fluorophore-conjugated secondary antibodies (*see* Table 1).

19. ProLong Gold anti fade mounting medium (Thermo Fisher Scientific, MA, USA).

20. Zeiss 710.

21. Zeiss microscope objective lenses: 60 X.

22. Zen Blue Software for image acquisition and analysis.

3   Method

1. Warm RPMI with 5% FBS at 37 °C before use.

2. Purify B cells from mouse spleen. Resuspend purified B cells in RPMI containing 5% FBS.

3. Plate approximately 0.5 million B cells in 300 μL medium per well.

4. Cover the chamber slides with the provided lid. Incubate at 37 °C in a humidified incubator with 5% CO_2 for 20 min to allow attachment of cells (*see* **Note 3**).

5. Remove chamber slides, tilt the slides slightly, and carefully aspirate the medium completely by placing the pipette tip at the corner. Make sure that the pipette tip does not touch the cells (*see* **Note 4**).

6. Quickly add 250 μL warm RPMI containing 2% FBS from the side of the wells (*see* **Notes 5** and **6**). Do not directly pipette in the middle (*see* **Note 7**).

7. Stimulate cells with either 10 μg/mL Alexa 647 labeled goat F(ab)$_2$ anti-mouse IgM (Fig. 1) or 10 μg/mL unlabelled goat F(ab)$_2$ anti-mouse IgM (Fig. 2) (*see* **Note 8**). Keep one well unstimulated for detecting each signaling protein (*see* **Note 9**). Incubate at 37 °C for the indicated time. Keep samples that were stimulated with Alexa 647 labeled anti-IgM protected from light.

8. At the end of incubation, quickly aspirate stimulant-containing medium completely.

9. Wash two times with copious amount of ice-cold PBS.

10. Fix the cells with 350 μL of freshly prepared 3.7% (vol/vol) paraformaldehyde. Incubate for 20 min at RT (*see* **Note 10**).

11. Completely aspirate the fixative.

12. Wash three times for 5 min each with PBS with gentle agitation (50–55 RPM on a flat rotator).

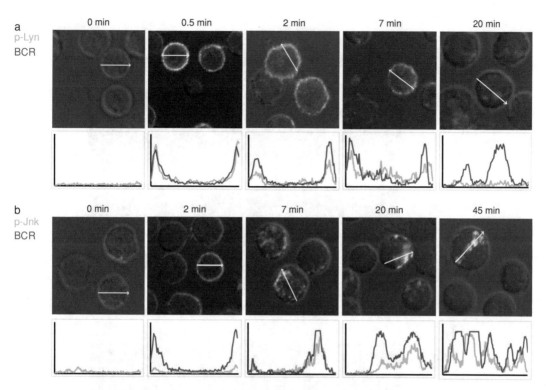

Fig. 1 Confocal images of phospho-kinases following BCR crosslinking. B cells were treated with Alexa 647-anti-IgM and stained with Abs specific for p-Lyn, or p-JNK followed by Alexa 488-conjugated secondary Abs. Confocal images of Alexa 488 (*green*) and Alexa 647 (*red*) merged with DIC images are shown. An intensity analysis of the p-kinase (*green*) and BCR (*red*) for representative cells in the plane of the image in the direction shown by the *white arrow* is given below each panel. The *x*-axis represents the distance of the *white arrow*, approximately 6 μm and the *y*-axis represents the fluorescent intensities of Alexa 488 (*green*) and Alexa 647 (*red*)

13. Remove PBS and add 350 μL freshly prepared permeabilization buffer (0.2% Triton X-100 in 1 X PBS) (*see* **Note 11**). Incubate at RT with gentle agitation for 30 min.

14. Remove permeabilization buffer, rinse once, and add blocking solution (1 X PBS, 0.2% Triton X-100, 2% BSA, 50 μg/mL normal goat serum). Incubate at RT with gentle agitation for 45 min.

15. While the slides are incubating, prepare working dilutions of primary antibodies in 1 X PBS, 0.2% Triton X-100, 2% BSA (*see* Table 1).

16. After blocking is complete, aspirate buffer and add 200 μL of primary antibodies solution.

17. Allow the antibodies to bind to the cells by incubating for either 4 h at RT or overnight at 4 °C. We prefer to incubate at 4 °C overnight without shaking and then 1 h at RT on the rotator at 55 RPM (*see* **Note 12**).

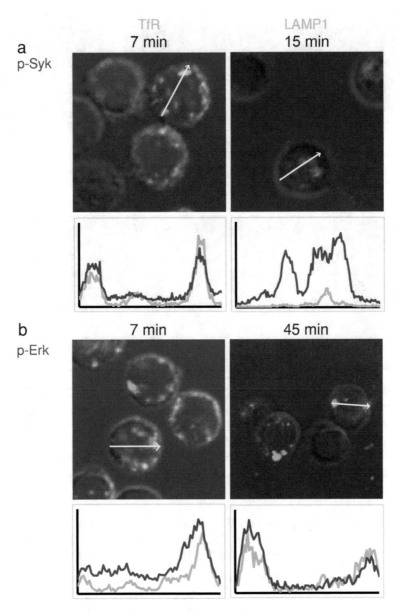

Fig. 2 Colocalization of p-kinases with TfR and LAMP1 following BCR crosslinking. B cells were incubated with unlabeled anti-IgM at 37 °C for the indicated times. (**a**) Confocal images merged with DIC images of cells stained with Abs specific TfR or LAMP1 followed by secondary Abs conjugated to Alexa 488 (*green*) and phospho-Syk followed by secondary Abs conjugated to Alexa 647 (*red*) are shown. (**b**) Confocal images merged with DIC images of cells stained with Abs-specific TfR or LAMP1 followed by secondary Abs conjugated to Alexa 488 (*green*) and phospho-Erk followed by secondary Abs conjugated to Alexa 647 (*red*) are shown. An intensity analysis of Alexa 488 (*green*) and Alexa 647 (*red*) in representative cells in the direction shown by the *white arrow* is given below each image. The *x*-axis represents the distance of the *white arrow*, approximately 6 μm, and the *y*-axis represents the fluorescent intensities of Alexa-488 (*green*) and Alexa 647 (*red*)

18. Wash three times for 10 min each with PBS at RT with gentle agitation.

19. Completely remove PBS.

20. While the slides are washing, prepare working dilutions of fluorophore-conjugated secondary antibodies in 1 X PBS, 0.2% Triton X-100, 2% BSA (*see* Table 1).

21. Add 200 μL secondary antibodies solution and incubate for 1 h at RT with gentle agitation. Keep all the slides protected from light to prevent attenuation of fluorophores.

22. Wash three times for 10 min each with PBS at RT with gentle agitation.

23. Aspirate all PBS completely by tilting the slides.

24. Carefully remove the chamber from the bottom glass slide by using the small tool that comes with BioCoat chamber slides. Insert the tool between chambers and glass slide as far as possible to prevent breaking of slides.

25. Tilt the slide and allow the edge of the slide to touch the paper towel so that any residual liquid can be absorbed. Do not allow paper towel to touch the sample directly.

26. Remove any residual adhesive from the border of the slide (*see* **Note 13**).

27. Add a small drop of mounting medium on each sample, invert the cover slip gently, lay it down on top of specimen, allowing the mounting medium to spread across the sample evenly, avoiding any air bubbles between the slides and the coverslips (*see* **Note 14**). Make sure that the cover glass does not move. Do not press the cover glass (*see* **Note 15**).

28. Wipe away excess mounting medium and completely seal all edges with nail polish.

29. Allow it to dry for atleast 3–4 h. Store mounted slides at 4 °C in the dark.

30. Slides can be stored at −20 °C for long-term storage.

31. Image slides using standard confocal procedures on a Zeiss 710 Meta confocal microscope with 60 X objective. Analyze images by Zen software. To get the intensities click on the profile tab. Select the arrow sign and draw an arrow from one end to another end of the cell. Save the image with an arrow as a tiff file and intensities as a text file. Plot a graph of intensities for the two fluorophores using Microsoft excel.

4   Notes

1. B cells can be purified as described earlier [19]. An alternative to using primary B cells is to use murine B cell lines.

2. BioCoat chamber slides should be stored at 4 °C. Because these slides are precoated, they do not require coating with Poly-L-Lysine. Alternatively, Labtec chambers, glass slides, or cover glasses can be used, which would require coating with poly-L-Lysine or Poly-D-Lysine.

3. Keep chamber slides covered with the lid during all incubations to prevent evaporation.

4. Aspirate the medium slowly. Attach a 200 μL pipette tip to the 2 mL pipette connected to a vacuum waste tank.

5. Keep the slides moist and do not dry them completely any time after this point until mounting step.

6. FBS can be replaced with 2% BSA. However, we prefer using FBS.

7. Pipette all reagents from the side of the wells without disturbing the attached cells. Directly pipetting the solution on the cells dislodges them.

8. We stimulate the cells after attaching them to the chamber slides. Alternatively, stimulation can be done in eppendorf tubes and following fixation cells can be attached to Poly-L-Lysine-coated glass slides. However, after fixation cells do not adhere strongly. Moreover, many cells will not be in the same plane for imaging.

9. Keep few wells under unstimulated condition in each chamber slide, for staining with primary antibody of interest, isotype control, and secondary antibody control.

10. Alternatively, methanol fixation can also be performed at −20 °C. However, we find that PFA works better for all the antibodies tested here.

11. Alternative detergents such as Saponin, SDS, or Tween 20 can also be used. We find that both Triton X-100 and Saponin work well.

12. For Fig. 2 we have incubated the cells with both primary antibodies (antibodies to pSyk or pERK and TfR or LAMP1) at the same time as they are raised in different hosts. Similarly both fluorophore labeled secondary antibodies were added at the same time. However, care must be taken at this step to ensure that the antibodies under co-incubation are not derived from the same host.

13. Take off any residual adhesive from the border of slide. It increases the distance between the coverslip and the cells and interferes with image acquisition.

14. Vectashield mounting medium (Vector laboratories) can also be used. We do not use DAPI because of the very large nucleus and relatively smaller cytoplasm of B cells.

15. 24 × 50 mm cover glass fits perfectly on the BioCoat slide.

Acknowledgments

A.R.M. is the recipient of DST-INSPIRE fellowship. A.C. is supported by intramural funds by IISER Pune.

References

1. Cambier JC, Gauld SB, Merrell KT, Vilen BJ (2007) B-cell anergy: from transgenic models to naturally occurring anergic B cells? Nat Rev Immunol 7:633–643

2. Reth M, Wienands J (1997) Initiation and processing of signals from the B cell antigen receptor. Annu Rev Immunol 15:453–479

3. Dalporto J (2004) B cell antigen receptor signaling 101. Mol Immunol 41:599–613

4. Campbell KS (1999) Signal transduction from the B cell antigen-receptor. Curr Opin Immunol 11:256–264

5. Kurosaki T, Shinohara H, Baba Y (2010) B cell signaling and fate decision. Annu Rev Immunol 28:21–55

6. Clark MR, Massenburg D, Zhang M, Siemasko K (2003) Molecular mechanisms of B cell antigen receptor trafficking. Ann N Y Acad Sci 987:26–37

7. Lanzavecchia A (1990) Receptor-mediated antigen uptake and its effect on antigen presentation to class II-restricted T lymphocytes. Annu Rev Immunol 8:773–793

8. Chaturvedi A, Martz R, Dorward D, Waisberg M, Pierce SK (2011) Endocytosed BCRs sequentially regulate MAPK and Akt signaling pathways from intracellular compartments. Nat Immunol 12:1119–1126

9. Murphy JE, Padilla BE, Hasdemir B, Cottrell GS, Bunnett NW (2009) Endosomes: a legitimate platform for the signaling train. Proc Natl Acad Sci U S A 106:17615–17622

10. Sadowski L, Pilecka I, Miaczynska M (2009) Signaling from endosomes: location makes a difference. Exp Cell Res 315:1601–1609

11. Vieira AV, Lamaze C, Schmid SL (1996) Control of EGF receptor signaling by clathrin-mediated endocytosis. Science 274:2086–2089

12. Nazarewicz RR, Salazar G, Patrushev N, Martin AS, Hilenski L, Xiong S, Alexander RW (2011) Early endosomal antigen 1 (EEA1) is an obligate scaffold for angiotensin II-induced, PKC- -dependent Akt activation in endosomes. J Biol Chem 286:2886–2895

13. Chaturvedi A, Pierce SK (2009) How location governs toll-like receptor signaling. Traffic 10 (6):621–628

14. Chaturvedi A, Dorward D, Pierce SK (2008) The B cell receptor governs the subcellular location of Toll-like receptor 9 leading to hyperresponses to DNA-containing antigens. Immunity 28:799–809

15. Delcroix J-D, Valletta JS, Wu C, Hunt SJ, Kowal AS, Mobley WC (2003) NGF signaling in sensory neurons: evidence that early endosomes carry NGF retrograde signals. Neuron 39:69–84

16. Wu C, Lai CF, Mobley WC (2001) Nerve growth factor activates persistent Rap1 signaling in endosomes. J Neurosci 21:5406–5416

17. Omerovic J, Prior IA (2009) Compartmentalized signalling: Ras proteins and signalling nanoclusters. FEBS J 276:1817–1825

18. Taelman VF, Dobrowolski R, Plouhinec J-L, Fuentealba LC, Vorwald PP, Gumper I, Sabatini DD, De Robertis EM (2010) Wnt signaling requires sequestration of glycogen synthase kinase 3 inside multivesicular endosomes. Cell 143:1136–1148

19. Chaturvedi A, Siddiqui Z, Bayiroglu F, Rao KVS (2002) A GPI-linked isoform of the IgD receptor regulates resting B cell activation. Nat Immunol 3:951–957

Chapter 10

Imaging the Interactions Between B Cells and Antigen-Presenting Cells

Jia C. Wang, Madison Bolger-Munro, and Michael R. Gold

Abstract

In vivo, B cells are often activated by antigens that are displayed on the surface of antigen-presenting cells (APCs). Binding of membrane-associated antigens to the B cell receptor (BCR) causes rapid cytoskeleton-dependent changes in the spatial organization of the BCR and other B cell membrane proteins, leading to the formation of an immune synapse. This process has been modeled using antigens attached to artificial planar lipid bilayers or to plasma membrane sheets. As a more physiological system for studying B cell-APC interactions, we have expressed model antigens in easily transfected adherent cell lines such as Cos-7 cells. The model antigens that we have used are a transmembrane form of a single-chain anti-Igκ antibody and a transmembrane form of hen egg lysozyme that is fused to a fluorescent protein. This has allowed us to study multiple aspects of B cell immune synapse formation including cytoskeletal reorganization, BCR microcluster coalescence, BCR-mediated antigen gathering, and BCR signaling. Here, we provide protocols for expressing these model antigens on the surface of Cos-7 cells, transfecting B cells with siRNAs or with plasmids encoding fluorescent proteins, using fixed cell and live cell fluorescence microscopy to image B cell-APC interactions, and quantifying APC-induced changes in BCR spatial organization and signaling.

Key words B cells, B cell receptor, Antigen-presenting cell, Immune synapse, Confocal microscopy, Cytoskeleton, Signal transduction

1 Introduction

The binding of cognate antigen to the B cell receptor (BCR) initiates signaling cascades that regulate B cell activation. Mature B cells home to secondary lymphoid organs where antigens are delivered via the blood and lymphatic systems [1]. Within the lymphoid organ, follicular dendritic cells, dendritic cells, and subcapsular sinus macrophages capture antigens and present them to B cells [2–6]. Although B cells readily bind soluble antigens, in vivo B cells may often be activated by antigens that are displayed on the

---

Jia C. Wang and Madison Bolger-Munro are the co-first authors

Chaohong Liu (ed.), *B Cell Receptor Signaling: Methods and Protocols*, Methods in Molecular Biology, vol. 1707, https://doi.org/10.1007/978-1-4939-7474-0_10, © Springer Science+Business Media, LLC 2018

surface of antigen-presenting cells (APCs) [2]. Hence, there is considerable interest in understanding how membrane-bound antigens initiate B cell activation.

When B cells engage membrane-associated antigens, dynamic changes in cell morphology and membrane protein organization occur at the B cell-APC contact site [7–9]. Antigen binding rapidly causes BCRs to form microclusters that recruit signaling enzymes. Transient spreading of the B cell across the surface of the APC increases the number of antigen encounters. Subsequent contraction of the B cell membrane is accompanied by the coalescence of antigen-bound BCR microclusters into a central supramolecular activation cluster (cSMAC) that is surrounded by a ring of integrins and an underlying ring of F-actin [7, 10, 11]. This characteristic bulls-eye immune synapse pattern develops in ~10 min and can be visualized by the procedures described in this chapter. Immune synapse formation optimizes the two functions of the BCR, signal transduction, and antigen acquisition for the purpose of eliciting T cell help.

Membrane protein mobility and spatial organization are controlled by the cytoskeleton, which plays a critical role in immune synapse formation. In resting B cells, a network of submembrane F-actin restricts BCR mobility [12–14]. BCR signaling stimulates local breakdown and reorganization of the actin cytoskeleton, which increases BCR mobility and promotes microcluster formation. Remodeling of the actin cytoskeleton drives B cell spreading and contraction, as well as the coalescence of microclusters into a cSMAC [15]. F-actin reorganization at the immune synapse also supports BCR-induced polarization of the microtubule network toward the B cell-APC contact site [16]. Juxtamembrane microtubules at the B cell-APC contact site are required for BCR microclusters to coalesce into a cSMAC [17]. Reorientation of the microtubule network toward the immune synapse also promotes antigen acquisition and the delivery of antigens to antigen-processing and MHC-loading compartments [18]. Understanding how cytoskeletal reorganization at the B cell-APC contact site is controlled, and how this drives immune synapse formation, is critical for understanding APC-induced B cell activation.

Elucidating the mechanisms underlying immune synapse formation requires an in vitro system where microscopy can be used to reveal the spatiotemporal reorganization of the BCR, other membrane proteins that regulate BCR signaling (e.g., CD19 and CD22), cytoskeletal structures, and signaling proteins. Most studies have employed antigens that are immobilized on coverslips, attached to artificial planar lipid bilayers, or linked to plasma membrane sheets that are derived from mammalian cells. When B cells are allowed to settle onto these planar antigen arrays, the interface between the B cell membrane and the antigen-bearing substrate is in a single x-y focal plane, which can be imaged with high spatial and temporal resolution.

Antigens that are immobilized on a rigid 2-dimensional surface (e.g., a coverslip) are useful for studying some of the early events in immune synapse formation, particularly cell spreading. However, the immobility of the antigens limits the formation, growth, and lateral movement of BCR microclusters and prevents their coalescence into a cSMAC [19].

In contrast, antigens and adhesion molecules that are linked to artificial planar lipid bilayers display unrestrained lateral mobility and support the coalescence of BCR microclusters into a cSMAC [8, 10, 20, 21]. The ability to independently control the nature and densities of antigens, integrin ligands, and other molecules that are attached to the bilayers has enabled quantitative analysis of the relationship between receptor engagement, BCR signaling, and B cell activation by membrane-bound antigens [10, 11]. A caveat associated with the use of planar lipid bilayers is that they do not reflect the complexity of the APC membrane. Cell membranes are organized into domains with distinct lipid compositions; proteins are organized into nanoclusters or proteins islands; and the sub-membrane F-actin cytoskeleton restricts the lateral mobility of proteins within the membrane. Therefore, this system may not accurately recapitulate all of the molecular events that occur at the immune synapse. For example, the mobility of ICAM-1 within the APC membrane must be constrained by the actin cytoskeleton in order to support T cell activation [22]. Moreover, in contrast to biological membranes, B cells cannot readily extract and acquire antigens from artificial planar lipid bilayers [23, 24]. Hence, this important aspect of B cell immune synapse function cannot be studied using planar lipid bilayers.

To address these limitations, Tolar and colleagues have employed plasma membrane sheets generated by sonicating cells that have been adhered to coverslips [24, 25]. Antigens are attached to the membrane sheet via annexin-V-biotin-streptavidin tethers. These membrane sheets retain the complexity of the plasma membrane and some of the cortical actin cytoskeleton remains intact. However, the cytoplasmic side of the membrane faces upward and interacting B cells would abnormally encounter the intracellular domains of transmembrane proteins [25]. As well, the streptavidin tethers form small clusters and B cells may therefore encounter pre-clustered antigens. Nevertheless, this system supports cSMAC formation and has been useful for studying the mechanical forces involved in BCR-mediated antigen extraction [24].

The use of live, intact APCs represents the most physiological system for studying B cell-APC interactions. Follicular dendritic cells may be the most important APC for B cells in vivo, but these cells are present in very low numbers and are difficult to isolate [26]. We have taken the approach of expressing model antigens in easily transfected adherent cell lines (Fig. 1a). Thus far we have

A

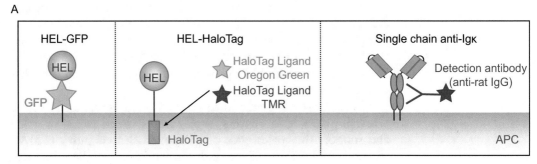

B

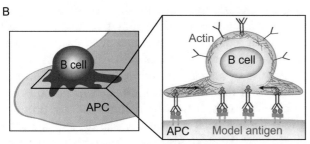

Fig. 1 Schematic of model antigens and APCs. (**a**) The model antigens that we have expressed on the surface of Cos-7 cells are shown. *Left*, HEL-GFP has the complete HEL and GFP proteins in the extracellular domain, which are fused to the transmembrane region and cytosolic domain of H-2K^b. *Center*, HEL-HaloTag has the complete HEL protein in the extracellular domain, which is joined to the transmembrane region and cytoplasmic domain of H-2K^b. The HaloTag protein has been fused to the C-terminus of the H-2K^b cytoplasmic domain. Different cell-permeable HaloTag ligands can be used to add a fluorescent tag to the fusion protein. *Right*, The single-chain transmembrane anti-Igκ antibody is a dimeric fusion protein consisting of the F$_v$ from the 187.1 rat anti-mouse Igκ monoclonal antibody fused to the juxtamembrane and transmembrane regions of rat IgG1. This fusion protein can be visualized using fluorescent anti-rat IgG antibodies. (**b**) *Left*, B cells (*red*) are allowed to settle onto antigen-bearing Cos-7 cells (*green*). *Right*, Binding of the model antigen to BCRs triggers microcluster formation and dynamic reorganization of the actin cytoskeleton

expressed two different model antigens, a dimeric transmembrane form of a single-chain anti-Igκ antibody [27] and a transmembrane form of hen egg lysozyme (HEL) that is fused to a fluorescent protein. We have expressed these proteins in Cos-7 African green monkey kidney cells or in B16F1 murine melanoma cells [15, 16, 28], both of which can be transfected with high efficiency. When plated on extracellular matrix components, these cells spread and flatten such that when B cells contact them, the B cell-APC interface is in a single *x-y* confocal plane. This system has allowed us to image immune synapse formation and identify both signaling proteins and cytoskeletal regulators that drive this process [15, 16, 28]. Cos-7 cells do not appear to express ligands for murine integrins or other receptors on murine B cells. Thus, Cos-7 cells expressing these model antigens can be used to investigate how BCR signaling regulates immune synapse formation in the absence of contributions by integrins or co-receptors. Co-expressing a model antigen along with an integrin ligand or other relevant APC proteins should be possible, allowing one to take a gain-of-function approach to engineering the APC. A caveat of this system is that it is difficult to control the antigen density on the APCs and the surface expression of the model antigen can vary from cell to cell. Nevertheless, this is a robust and powerful experimental system for imaging the interaction of B cells with live, intact APCs.

Here, we provide protocols for generating APCs expressing model antigens and for using fluorescence microscopy to visualize and quantify APC-induced changes in BCR spatial organization and signaling. Detailed protocols are provided for both fixed cell and live cell imaging of B cell-APC interactions. Fixed cell imaging allows one to use multiple probes to visualize the spatial organization of the BCR, cytoskeletal structures, and readouts of BCR signaling. Live cell imaging using fluorescent model antigens, together with genetically encoded fluorescent proteins or probes, provides superior temporal resolution of the events occurring at the B cell-APC contact site. These experimental systems can be used to obtain new insights into the mechanisms underlying B cell immune synapse formation and function.

2   Materials

2.1 Preparing Coverslips for Adhering APCs

1. 18 mm diameter round No. 1 Micro Cover Glasses, 0.13–0.17 mm thickness (VWR, catalogue #48380-046).

2. Forceps (tapered ultrafine tip, e.g., Fisher Scientific, catalogue #16100121).

3. 100% methanol.

4. Falcon 12-well sterile tissue polystyrene tissue culture plate (Corning, catalogue #353043).

5. Sterile phosphate-buffered saline (PBS) without calcium or magnesium.

6. Fibronectin from bovine plasma (Sigma-Aldrich, catalogue #F4759; BioReagent suitable for cell culture). Dissolve the fibronectin in sterile water to a final concentration of 5 μg/ml. Store at 4 °C.

2.2 Passaging Cos-7 Cells

1. Cos-7 cells: SV40-transformed African green monkey kidney cells (ATCC, catalogue #CRL-1651). Use biosafety level 2 procedures for handling these cells.

2. Complete DMEM medium: Dulbecco's Modified Eagle Medium (DMEM with 4.5 g/l glucose; Thermo Fisher, catalogue #11960069) supplemented with 10% fetal calf serum (FCS), 2 mM glutamine, 1 mM pyruvate, 50 U/ml penicillin, and 50 μg/ml streptomycin.

3. Falcon 10 cm sterile polystyrene tissue culture dish (Corning, catalogue #353003).

4. Sterile PBS.

5. Sterile trypsin-EDTA solution (0.25% trypsin, 1 mM EDTA; Thermo Fisher catalogue #25200-072).

2.3 Expressing
Model Antigens
in Cos-7 Cells
and Seeding the Cells
onto Coverslips

1. Falcon 6-well sterile polystyrene tissue culture plate (Corning, catalogue #353046).

2. Plasmid DNA encoding the model antigen. All plasmids are purified using the GeneJET Plasmid Mini-prep kit (Thermo Fisher, catalogue #K0503).

   (a) Transmembrane dimeric single-chain anti-Igκ antibody (Fig. 1a) [27]: This fusion protein consists of a single-chain Fv generated from the 187.1 rat anti-Igκ monoclonal antibody, the hinge and membrane-proximal domains of rat IgG1, and the transmembrane and cytoplasmic domains of H-2K^b. This plasmid is available upon request from the authors, with permission from the originator (Dr. David Nemazee, Scripps Institute, La Jolla, CA).

   (b) Transmembrane form of HEL fused to GFP (HEL-GFP; (Fig. 1a) [7]. This fusion protein consists of the complete wild-type HEL protein, the complete EGFP protein, the transmembrane region of H-2K^b, and the 23-amino acid cytoplasmic domain of H-2K^b. This plasmid is available upon request from the authors, with permission from the originator (Dr. Facundo Batista, Ragon Institute, Cambridge, MA).

   (c) Transmembrane form of HEL fused to HaloTag (HEL-HaloTag; Fig. 1a). This fusion protein consists of the complete wild-type HEL protein, the transmembrane region and 23-amino acid cytoplasmic domain of H-2K^b, and the complete HaloTag protein. The DNA fragment encoding the HaloTag protein was excised from the pHTC HaloTag CMV-neo vector (Promega, catalogue #G7711). To avoid excessively high levels of expression, the HEL-HaloTag construct was subcloned into the pβN1 expression vector, a derivative of pEGFP-C1 in which the EGFP gene was removed and the CMV promoter was replaced with the weaker β-actin promoter. This plasmid is available upon request from the authors.

3. Lipofectamine 2000 (Thermo Fisher, catalogue #11668019) or Lipofectamine 3000 transfection reagent (Thermo Fisher, catalogue # L3000008).

4. Opti-MEM (Thermo Fisher, catalogue #31985062) or DMEM (Thermo Fisher, catalogue #11960069).

5. Sterile PBS.

6. Enzyme-free dissociation buffer (0.5 mM EDTA, 100 mM NaCl, 1 mM glucose, pH 7.4). Add 0.584 g NaCl and 0.1 ml of 0.5 M EDTA to 90 ml distilled water. Adjust the pH to 7.4 and then adjust the volume to 100 ml. Sterile filter through a 0.22 μm filter and store at 4 °C.

2.4 Isolating Primary B Cells

1. C57BL/6 mouse or MD4 mouse [29] (Jackson Laboratories, catalogue #002595). All of the B cells in MD4 mice express a transgenic BCR specific for HEL. B cells from MD4 mice are used in conjunction with APCs expressing HEL-GFP or HEL-HaloTag.

2. Sterilized surgical scissors and forceps.

3. Falcon 70 μm cell strainer (Corning, catalogue #352350).

4. EasySep™ Mouse B cell Isolation kit (Stemcell Technologies, catalogue #19854).

5. EasySep™ Buffer (Stemcell Technologies, catalogue #20144).

6. Falcon 5 ml polystyrene 12 × 75 mm round-bottom tubes (Corning, catalogue #352058).

7. 50 ml centrifuge tube.

8. Modified HEPES-buffered saline with 2% FCS (mHBS-FCS): Add 2 ml of heat-inactivated FCS to 100 ml of mHBS (25 mM HEPES, pH 7.2, 125 mM NaCl, 5 mM KCl, 1 mM $CaCl_2$, 1 mM Na_2HPO_4, 0.5 mM $MgSO_4$, 1 mg/ml glucose, 2 mM glutamine, 1 mM sodium pyruvate, 50 μM 2-mercaptoethanol).

9. Complete RPMI medium: RPMI-1640 (Thermo Fisher, catalogue #21870-076) supplemented with 10% heat-inactivated FCS, 2 mM glutamine, 1 mM pyruvate, 50 μM 2-mercaptoethanol, 50 U/ml penicillin, and 50 μg/ml streptomycin.

10. *E. coli* 0111:B4 lipopolysaccharide (LPS; Sigma-Aldrich, catalogue #L4391).

11. Recombinant mouse B cell-activating factor (BAFF; R&D Systems, catalogue #8876-BF-010).

12. FACS buffer (PBS with 2% heat-inactivated FCS and 0.1% sodium azide).

13. FITC-conjugated goat anti-mouse IgM (Jackson ImmunoResearch, catalogue #115-095-075).

2.5 Culturing A20 and A20/D1.3 Cells

1. A20 cells: murine B cell lymphoma of Balb/c origin that expresses an IgG-containing BCR on its surface (ATCC, catalogue #TIB-208).

2. A20/D1.3 cells: A20 cells that express a transgenic HEL-specific IgM BCR on their surface in addition to the endogenous IgG BCR [30]. These cells are available upon request from the authors, with permission from the originator (Dr. Facundo Batista, Ragon Institute, Cambridge, MA).

2.6 Transfecting Primary B Cells or A20 Cells with siRNAs or Plasmid DNA

1. *E. coli* 0111:B4 lipopolysaccharide (LPS; Sigma-Aldrich, catalogue #L4391).

2. Mouse recombinant B cell-activating factor (BAFF; R&D Systems, catalogue #8876-BF-010).

3. siRNA pools for the target protein (e.g., from GE Dharmacon) as well as a control siRNA pool (e.g., ON-TARGETplus Non-Targeting Pool, GE Dharmacon, catalogue #D-001810-01-05).

4. Plasmid encoding mTagRFP-α-tubulin (Addgene, #58026) or GFP-α-tubulin (Addgene, catalogue #56450).

5. Plasmid encoding LifeAct-GFP (Addgene, catalogue #58470), mTagRFP-T-Lifeact-7 (Addgene, catalogue #54586), LifeAct-mCherry (Addgene, catalogue #54491), or F-tractin-tdTomato. The F-tractin-tdTomato plasmid [16, 31] can be requested from Dr. John Hammer (National Heart, Lung, and Blood Institute, NIH, Bethesda, MD).

6. Amaxa Nucleofection kit V (Lonza, catalogue #VCA-1003).

7. Amaxa Nucleofector™ model 2b (Lonza, catalogue #AAB-1001).

2.7 B Cell-APC Interactions Followed by Cell Fixation

1. Latrunculin A (Enzo Life Sciences, catalogue #BML-T119). Dissolve in DMSO to a final concentration of 1 mM. Store at −20 °C protected from light.

2. Nocodazole (Sigma-Aldrich, catalogue #M1404). Dissolve in DMSO to a final concentration of 1 mM. Store at 4 °C protected from light.

3. 4 or 8% paraformaldehyde (PFA): Dilute 1 ampoule of 10 ml aqueous solution of 16% PFA (Electron Microscopy Sciences, catalogue #15710) with PBS to achieve the desired working concentrations. Store at room temperature.

2.8 Permeabilization and Staining of B Cell-APC Conjugates on Coverslips

1. 0.2% (w/v) Triton X-100: Dilute 10% (w/v) Triton X-100 stock 1:50 with PBS.

2. Blocking buffer (PBS containing 2% bovine serum albumin (BSA) and 0.1% Triton X-100): Dissolve 2 g BSA in 90 ml PBS, add 1 ml of 10% (w/v) Triton X-100, and adjust the volume to 100 ml with PBS. Sterile filter and store at 4 °C. Warm to room temperature before use.

3. Primary antibodies that we have used include rabbit anti-phosphotyrosine (P-Tyr; Cell Signaling Technology catalogue

#8954), rabbit anti-phospho-CD79a (P-CD79a; Cell Signaling Technology catalogue #5173), and rat anti-tubulin (Abcam, catalogue #ab6161).

4. For detecting the single-chain anti-Igκ on the APCs, use goat anti-rat IgG (Fc region-specific, no cross-reactivity with mouse IgG) conjugated to Alexa Fluor 488 (Thermo Fisher, catalogue #A11006) or Alexa Fluor 647 (Thermo Fisher, catalogue #A21247).

5. Secondary antibodies that we have used (all from Thermo Fisher) include Alexa Fluor 488-conjugated goat anti-rabbit IgG (#A11034), Alexa Fluor 568-conjugated goat anti-rabbit IgG (#A11036), Alexa Fluor 568-conjugated goat anti-mouse IgG (#A11031), and Alexa Fluor 647-conjugated goat anti-mouse IgG (#A21235).

6. Phalloidin coupled to rhodamine, Alexa Fluor 568, or Alexa Fluor 647 (Thermo Fisher, catalogue #R415, #A12380, and #A22287, respectively).

7. ProLong Diamond antifade mounting reagent with DAPI (Thermo Fisher, catalogue #P36962) or without DAPI (Thermo Fisher, catalogue #P36961).

2.9   Imaging Fixed Cells Using a Laser Scanning Confocal Microscope

1. Any laser scanning confocal microscope system with a $100\times$ objective and the appropriate excitation lasers and emission filters is suitable for imaging B cell-APC interactions. The system we have used consists of the following:
   (a) Olympus FV1000 laser scanning confocal microscope system based on an IX81 inverted microscope.
   (b) UplanSApo $100\times/1.40$ NA oil objective.
   (c) Multi-line argon laser.
   (d) Fluoview version 4.1 (Olympus).

2.10   Preparing APCs and B Cells for Live Cell Imaging

1. mHBS-FCS (*see* Subheading 2.4).

2. HaloTag tetramethylrhodamine (TMR) ligand (Promega, catalogue #G8252) or HaloTag Oregon Green ligand (Promega, catalogue #G2802).

2.10.1   Labeling HEL-HaloTag-Expressing Cos-7 Cells with a HaloTag Ligand

2.10.2   Staining B Cells with a Membrane Dye

1. CellMask Deep Red Plasma Membrane Stain (Thermo Fisher, catalogue #C10046) or CELLVUE Maroon (Polysciences, catalogue #24847-1).

2. PBS with 2% heat-inactivated FCS (PBS-FCS).

3. mHBS-FCS (*see* Subheading 2.4).

2.10.3 Treating B Cells with Cytoskeletal Inhibitors	1. Latrunculin A (Enzo Life Sciences, catalogue #BML-T119). Dissolve in DMSO to a final concentration of 1 mM. Store at −20 °C protected from light.

2. Nocodazole (Sigma-Aldrich, catalogue #M1404). Dissolve in DMSO to a final concentration of 1 mM. Store at 4 °C protected from light.

2.11 B Cell-APC Interactions: Real-Time Imaging of Antigen (HEL-GFP or HEL-HaloTag) Cluster Formation Using Spinning Disk Microscopy

1. Chamlide 35 mm magnetic imaging chamber (Quorum Technologies, catalogue #CM-B18-1).

2. mHBS-FCS (*see* Subheading 2.4).

3. Spinning disk microscopy system consisting of the following components:

   (a) Axiovert 200M microscope (Zeiss).

   (b) 100×/1.46 NA Oil Plan-Fluor objective lens (Zeiss).

   (c) Microscope stage equipped with a universal mounting frame K (Zeiss, catalogue # 451352-0000-000).

   (d) Microscope heating element.

   (e) Acrylic cage incubator.

   (f) QuantEM 512SC camera (Photometrics).

   (g) Slidebook 6.0 software (Intelligent Imaging Innovations).

2.12 General Equipment

1. Tissue culture hood.

2. 37 °C tissue culture incubator with 5% CO_2 in air mixture.

3. Table-top centrifuge for pelleting cells, e.g., Beckman Coulter Allegra X-14R centrifuge.

4. Hemacytometer or other type of cell counter.

5. 37 °C water bath.

6. ImageJ software (https://imagej.nih.gov/ij/index.html).

7. Microsoft Excel.

---

3   Methods

3.1 Preparing Coverslips for Adhering APCs

Fibronectin-coated coverslips can be prepared in advance and stored at 4 °C for up to 1 week before adhering APCs to them.

1. In the tissue culture hood, use forceps to dip 18 mm coverslips into 100% methanol. Lean the coverslips at a 45° angle against the wall of a 12-well tissue culture plate (Fig. 2, left panel) and allow them to air dry completely for 30 min.

2. Lay the dried coverslips flat in the centers of the wells of the 12-well plate such that the coverslips do not touch the sides of the wells.

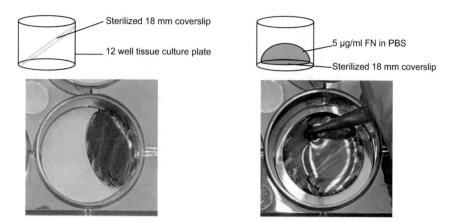

Fig. 2 Preparation of fibronectin-coated coverslips. *Left*, A methanol-rinsed 18 mm coverslip is leaned against the wall of a 12-well plate to be air-dried. *Right*, A dried coverslip is laid flat in the bottom of the well and covered with a fibronectin solution, forming a bubble on the top of the coverslip that is maintained by surface tension. A pipette tip is used to spread the fibronectin solution so that it covers the entire area of the coverslip. Note that the coverslip was colored red for clarity

3. Coat the coverslips with fibronectin. Pipet 370 μl of fibronectin (5 μg/ml in PBS) onto the center of the coverslip.

4. Use a sterile P-200 pipette tip to drag the fibronectin solution toward the edges of the coverslip so that it covers the entire coverslip as a rounded bubble (Fig. 2, right panel). Avoid letting the fibronectin solution contact the tissue culture plastic. The surface tension of the fibronectin solution should confine it to the coverslip.

5. Incubate the coverslips at room temperature for 30–60 min in the tissue culture hood. Avoid moving or vibrating the 12-well plate to ensure that the bubble of liquid does not disperse from the coverslip and flow into the rest of the well.

6. Wash the coverslip three times by filling each well with 1 ml of sterile PBS and then aspirating off the PBS. The coverslip is now ready to be seeded with the cells.

7. To store the coverslips before seeding them with the cells, add 1 ml of sterile PBS into each well containing a fibronectin-coated coverslip. Put the cover on the 12-well plate and store the coverslips at 4 °C for up to 1 week.

3.2 Passaging Cos-7 Cells

Before passaging the cells, pre-warm the tissue culture medium, sterile PBS, and trypsin-EDTA solution to 37 °C.

1. Culture Cos-7 cells in complete DMEM medium (*see* Subheading 2.2) in a 10 cm tissue culture dish until ~90% confluent.

2. In the tissue culture hood, wash the Cos-7 cells once with 5 ml of sterile PBS.

3. Aspirate off the PBS and add 1 ml of sterile trypsin-EDTA solution to the dish.

4. Place the dish in the 37 °C tissue culture incubator for 2 min.

5. Add 5 ml of complete DMEM medium to the dish and then pipette the liquid up and down three times with the end of the pipette pressed firmly against the bottom of the dish. The resulting turbulence helps dissociate cell clusters. Add 0.5–1 ml of this cell suspension to 9 ml of complete DMEM medium in a new 10 cm dish.

3.3 Expressing Model Antigens in Cos-7 Cells and Seeding the Cells onto Coverslips

Begin preparing the APCs 3 days before you wish to image B cell-APC interactions. As in Subheading 3.2, grow the Cos-7 cells in a 10 cm tissue culture dish until ~90% confluent and then detach the cells using trypsin-EDTA.

1. Transfer the Cos-7 cells into a sterile 15 ml conical tube and centrifuge at 1500 rpm ($525 \times g$) for 5 min at room temperature. Aspirate the supernatant and resuspend the cells in 5 ml of complete DMEM medium.

2. Count the cells using a hemacytometer or other types of cell counter.

3. Add 4.5×10^5 Cos-7 cells per well to a 6-well tissue culture plate. Add complete DMEM medium such that the final volume in each well is 2 ml. One well of Cos-7 cells is sufficient to seed 12 of the 18 mm coverslips.

4. Culture the cells overnight at 37 °C in the tissue culture incubator. After 18–24 h the Cos-7 cells should be 70–90% confluent.

5. Use Lipofectamine 2000 or Lipofectamine 3000 according to the manufacturer's protocol to transfect the Cos-7 cells with plasmid DNA encoding the single-chain anti-Igκ antibody, the transmembrane HEL-GFP, or the transmembrane HEL-HaloTag model antigens. Lipofectamine transfections are briefly described below. The volumes indicated are for one well of Cos-7 cells in a 6-well tissue culture plate. For one transfection reaction we generally use 2–2.5 μg of the single-chain anti-Igκ or transmembrane HEL-GFP plasmids or 0.25 μg of the transmembrane HEL-HaloTag plasmid. These amounts can be adjusted to achieve the desired level of antigen expression.

(a) Lipofectamine 2000: Dilute 10 μl of Lipofectamine reagent in Opti-MEM or DMEM to a final volume of 150 μl. In a separate tube, dilute the plasmid DNA (plasmid stock solutions are usually 0.5–5 μg/μl) in Opti-MEM or DMEM to a final volume of 150 μl; incubate at room temperature for 5 min. Mix the diluted

Lipofectamine reagent with the diluted DNA and incubate at room temperature for an additional 5 min.

(b) Lipofectamine 3000: Dilute 5 μl of Lipofectamine reagent in Opti-MEM or DMEM to a final volume of 125 μl. In a separate tube mix 5 μl of P3000 reagent with 120 μl Opti-MEM. Dilute plasmid DNA (plasmid stock solutions are usually 0.5–5 μg/μl) in the Opti-MEM/P3000 reagent mixture or DMEM/P3000 reagent mixture to a final volume of 125 μl. Mix the diluted Lipofectamine reagent with the diluted DNA and incubate at room temperature for 10–15 min.

(c) Using a P-1000 micropipette, add the entire DNA-lipid complex mixture dropwise to one well of the 6-well plate that contains Cos-7 cells in 2 ml of complete DMEM medium.

6. Culture the Cos-7 cells for 18–24 h in the tissue culture incubator.

7. Aspirate the medium and wash the Cos-7 cells twice with sterile PBS that has been pre-warmed to 37 °C.

8. Aspirate off the PBS and then gently pipette 500 μl of enzyme-free dissociation buffer (*see* Subheading 2.3) against the edge of the well and onto the cells. Place the cells in the 37 °C tissue culture incubator for 2 min.

9. Resuspend the cells by pipetting in 3 ml of complete DMEM medium. Pipette up and down three times to dislodge any remaining cells that are stuck to the bottom of the well. Transfer the cells into a 15 ml conical tube and centrifuge at 1500 rpm (525 × g) for 5 min at room temperature.

10. Aspirate the supernatant and resuspend the cells in 2 ml of complete DMEM medium by pipetting vigorously.

11. Count the cells and dilute to 1.5 × 10^5 cells per ml in a complete DMEM medium.

12. Rinse the fibronectin-coated coverslips that are in the 12-well plate with room temperature PBS (if the 12-well plate has been stored at 4 °C, warm it to room temperature before rinsing with PBS and adding Cos-7 cells to the wells). Add 1 ml (1.5 × 10^5 cells) of the transfected Cos-7 cells to each well of the 12-well tissue culture plate that contains a fibronectin-coated coverslip.

13. Place the cells in a 37 °C tissue culture incubator for at least 4 h, or overnight, to allow the cells to adhere to the coverslips, spread, and flatten. This improves the imaging of the B cell-APC interface since the contact site between the two cells will be in a single *x-y* confocal plane.

3.4 Isolating Primary B Cells

Isolate primary B cells early in the day on the day of the experiment. If transfecting the B cells with siRNAs, isolate the B cells 1 day prior to when you will add them to the APCs. The following protocol is for isolating primary B cells using the EasySep Mouse B Cell Isolation kit. A typical spleen from a 6–12 week old C57BL/6 mouse yields ~5 × 10^7 B cells. A spleen from an MD4 mouse will yield ~2 × 10^7 B cells.

1. Remove the spleen from a 6–12 week old C57BL/6 or MD4 mouse using sterilized surgical tools.

2. In the tissue culture hood, place a sterile 70 μm cell strainer into a 35 mm tissue culture dish containing 3 ml of sterile room temperature PBS. Use the rubber part of the plunger from a 5 ml syringe to mash the spleen through the cell strainer.

3. Pipette the PBS containing the single-cell suspension of splenocytes into a sterile 15 ml conical tube and centrifuge at 1500 rpm (525 × g) for 5 min at 4 °C.

4. Aspirate off the supernatant and isolate B cells using the EasySep Mouse B cell kit according to the manufacturer's instructions. A method that is adapted from the manufacturer's protocol is described below. All the steps must be performed at room temperature (15–25 °C).

   (a) Resuspend the pelleted splenocytes in 1 ml of EasySep buffer and transfer to a 5 ml 12 × 75 mm polystyrene tube that will fit into the EasySep magnet.

   (b) Add 60 μl of normal rat serum (included in the EasySep B cell isolation kit) to the splenocyte suspension and mix well by gently pipetting up and down.

   (c) Add 60 μl of EasySep mouse B cell Isolation Cocktail (antibody cocktail) to the splenocytes, mix gently by pipetting, and incubate for 10 min.

   (d) Vortex the EasySep Streptavidin RapidSpheres 50001 for 30 s to ensure that no aggregates are present. Add 90 μl of the RapidSpheres to the splenocytes, mix gently by pipetting, and incubate for 2.5 min.

   (e) Add 1.3 ml of EasySep Buffer to the splenocytes and mix gently by pipetting up and down twice. Remove the cap from the 5 ml tube containing the cells and place the tube in the EasySep magnet. Let the tube sit for 2.5 min.

   (f) Pick up the EasySep magnet with the tube still inside it and in one quick, continuous motion pour the contents from the 5 ml tube into a new 50 ml centrifuge tube. Do not remove the 5 ml tube from the magnet before pouring out the liquid and do not shake off drops that remain on the mouth of the tube.

5. If the primary B cells are to be used immediately for B cell-APC interaction experiments, centrifuge the cells at 1500 rpm ($525 \times g$) for 5 min at 4 °C, resuspend them to 10^6 per ml in modified HEPES-buffered saline containing 2% FCS (mHBS-FCS; *see* Subheading 2.4). Keep the cells on ice until you are ready to add them to the APCs.

    (a) If the primary B cells are to be transfected, resuspend the cells to 3×10^6 per ml in complete RPMI medium supplemented with 5 μg/ml LPS and 5 ng/ml BAFF and follow the steps described in Subheading 3.6.

6. Optional: To assess the purity of the B cell population, add 10 μl of cells to 90 μl of fluorophore-conjugated goat anti-mouse IgM that has been diluted to the appropriate concentration in ice-cold FACS buffer (*see* Subheading 2.4). Store the sample in the dark at 4 °C if the analysis will be performed later in the day. Use flow cytometry to determine the percent of live cells that are IgM$^+$. This procedure generally yields cell populations that are 90–95% IgM$^+$.

3.5 Culturing A20 and A20/D1.3 Cells

A20 B lymphoma cells grow as a mixture of suspension cells and lightly adherent cells.

1. Grow the cells in tissue culture plates or flasks in complete RPMI medium (*see* Subheading 2.4). Culture the cells at 37 °C in the tissue culture incubator (*see* Subheading 2.12).

2. Maintain the cells at a density of 10^5 to 10^6 cells per ml.

3. When passaging the cells, be sure to pipette the medium vigorously all around the bottom of the dish or flask to detach any adherent cells.

3.6 Transfecting Primary B Cells or A20 Cells with siRNAs or Plasmid DNA

If B cells are to be transfected with plasmid DNA or siRNAs, do so 1 day before using them to study B cell-APC interactions. Primary B cells should be isolated early on the day of transfection. Because naive primary B cells are difficult to transfect, we activate them with LPS for 6 h prior to transfection and then culture them overnight with the survival cytokine BAFF (in the absence of LPS) before adding them to APCs. For each transfection reaction, start with 3×10^6 LPS-activated primary B cells or 2.5×10^6 A20 cells. One transfection reaction provides enough B cells for 12 coverslips on which APCs have been grown. We use the Amaxa nucleofector electroporation system with Amaxa kit V to transfect both LPS-activated primary B cells and A20 cells. The Amaxa mouse B cell Nucleofector kit (catalogue #VPA-1010) can also be used to transfect LPS-activated primary B cells. To deplete specific proteins we have utilized commercial siRNA pools, as well as a combination of siRNAs targeting two different proteins. A pool of scrambled, non-targeting siRNAs should be used as a control. The extent of protein depletion should be monitored by immunoblotting. We have achieved substantial depletion of multiple proteins using this approach [16]. For live cell imaging, we have transfected B cells with fluorescent fusion proteins including LifeAct-GFP, LifeAct-mCherry, F-tractin-GFP, and F-tractin-tdTomato [31] to visualize actin dynamics, and with GFP-α-tubulin or mTagRFP-α-tubulin to visualize microtubule dynamics. As judged by the expression of fluorescent proteins, we have obtained transfection frequencies of 50–60% for A20 cells. The frequency of successfully transfecting a single cell with two plasmids, and obtaining similar expression levels for the two proteins, is much lower (~5%).

1. Prior to nucleofection, resuspend primary B cells to 3×10^6 per ml in complete RPMI medium supplemented with 5 μg/ml LPS and 5 ng/ml BAFF and culture the cells for 6 h at 37 °C. Then pellet the cells, resuspend them in complete RPMI medium (*see* Subheading 2.4) supplemented with 5 ng/ml BAFF, and culture the cells overnight at 37 °C.

2. For each transfection reaction, centrifuge 3×10^6 LPS-activated primary B cells or 2.5×10^6 A20 cells at 1500 rpm (525 $\times g$) for 5 min at room temperature.

3. Prepare the Amaxa Nucleofection Reagent. For each B cell sample that is to be transfected, mix 82 μl of Nucleofector Solution and 18 μl of Nucleofector supplement (~4.5:1 ratio) to yield a final volume of 100 μl of Nucleofection Reagent. A master mix can be prepared if nucleofecting multiple samples.

4. Aspirate the supernatant from the pelleted cells and resuspend the cells in 100 μl of the Nucleofection Reagent.

5. Add the siRNA or plasmid DNA to the cell mixture. Mix gently by pipetting up and down but avoid generating bubbles.

   (a) The amount of siRNA used must be optimized for the targeted protein. We use 0.5–2 μg of siRNA per reaction. Preliminary titration experiments should be performed to determine the amount of siRNA that yields the greatest knockdown with the least cell toxicity.

   (b) We use 2 μg of plasmid DNA per reaction. If transfecting with more than one plasmid, use 2 μg of each plasmid per reaction. The total volume of plasmid DNA added to the 100 μl of cell suspension should be as small as possible.

6. Transfer the mixture to an Amaxa nucleoporation cuvette, being careful to avoid bubbles. If bubbles develop, tap the bottom of the cuvette gently against the bench until the bubbles dissipate.

7. Nucleoporate the B cells using the Amaxa nucleoporator.

   (a) For A20 B cells use program L-013.

   (b) For LPS-activated primary B cells use program X-001.

8. Immediately after nucleofection, use the supplied transfer pipette to transfer the cells from the cuvette to a 6-well dish containing 2 ml of pre-warmed complete RPMI. Use an additional 500 µl of complete RPMI medium to wash out the cuvette and add this wash to the same well of the 6-well plate. For primary B cells, add 5 ng/ml of BAFF to the complete RPMI medium to promote cell survival.

9. Culture the cells for 18–48 h at 37 °C. Preliminary experiments should be performed to determine the amount of time required for optimal siRNA-mediated knockdown or optimal expression of transfected proteins.

10. To assess the extent of siRNA-mediated knockdown, at the end of this culture period remove 0.5 ml of the cells (~5 × 10^5 cells) and compare the expression of the target protein in untransfected cells, cells transfected with control (scrambled, non-targeting) siRNA, and cells transfected with siRNA specific for the protein of interest.

3.7 B Cell-APC Interactions Followed by Cell Fixation

The following procedures are used when B cells are added to Cos-7 APCs expressing the single-chain anti Igκ antibody on their surface. After fixation, B cell-APC conjugates are stained with fluorescent antibodies or other fluorescent probes (e.g., phalloidin to visualize F-actin). This protocol can also be used when HEL-specific B cells are added to HEL-GFP- or HEL-HaloTag-expressing APCs and you wish to visualize other components in addition to the model antigen.

1. Centrifuge B cells at 1500 rpm (525 × *g*) for 5 min at room temperature. Resuspend the cells to 10^6 per ml in mHBS-FCS (*see* Subheading 2.4) that has been pre-warmed to 37 °C.

2. At this point you can divide the B cells into different aliquots and pretreat them with chemical inhibitors (e.g., latrunculin A to depolymerize F-actin, nocodazole to depolymerize microtubules, signaling enzyme inhibitors). Control samples should receive equal volumes of the relevant solvent (e.g., DMSO). Inhibitor concentrations and pretreatment times should be optimized and validated.

3. Using a P-200 tip attached to a vacuum line, aspirate the medium from the wells that contain the coverslips on which Cos-7 cells have been grown. Wash the coverslips two times by carefully pipetting 500 µl of room temperature PBS against the side of the well (i.e., do not touch the coverslip) and then aspirating the PBS by gently tilting the plate and placing the P-200 tip against the side of the well. Add 37 °C mHBS-FCS to the wells.

4. Using one of the two methods described in **steps 5** and **6**, add the B cells to the APCs on the coverslip, allow the B cells to interact with the APCs for various lengths of time (see note below on time course), and then fix the cells with PFA.

   (a) Time course for visualizing immune synapse formation: BCR microcluster formation can be observed within 3 min of adding the B cells to the APCs and the coalescence of BCR microclusters into a cSMAC is evident by 10–15 min (*see* Fig. 4 as an example). By 30 min, immune synapse formation is complete and some of the BCR-bound antigen at the cSMAC may have been internalized.

5. Method 1 for B cell-APC interactions and cell fixation:

   (a) Cut a piece of parafilm so that it fits within a 15 cm diameter Petri dish. Write numbers on the parafilm that you will use to identify each coverslip. Place a Kimwipe that has been dampened with sterile water on the bottom of the Petri dish. Lay the parafilm on the top of the Kimwipe, ensuring that the parafilm is as flat as possible.

   (b) Using forceps, remove the coverslips from the wells and place them cell-side up on the parafilm. Be careful not to break the coverslips when removing them from the wells and only touch the forceps to the edge of the coverslip. Using a P-200 micropipette, add 100 μl of 37 °C mHBS-FCS to the coverslips so the cells do not dry out.

   (c) When you are ready to start the experiment, carefully aspirate off the liquid on the coverslip. Using a P-200 micropipette, pipette 100 μl of the B cell suspension (10^5 B cells) onto the edge of each coverslip so as not to disturb the monolayer of Cos-7 cells. If the B cell suspension does not spread completely over the coverslip, gently tilt the plate until the entire coverslip is covered.

   (d) Cover the dish with a lid and carefully place the dish with the coverslips in the 37 °C incubator.

   (e) After the B cells have been allowed to settle on the APCs and interact for the desired lengths of time, stop the reactions by fixing the cells. Add 100 μl of 8% PFA to the edge of the coverslip, being careful not to touch the coverslip. Minimize any turbulence that might disrupt B cell-APC interactions. Fix the cells at room temperature for 10 min.

6. Method 2 for B cell-APC interactions and cell fixation: Some users may find it easier to carry out the B cell-APC interaction and fixation steps with the coverslips in the wells of the 12-well plate. This method has been used in [15, 16, 28].

(a) After washing the coverslips as in **step 3** above, aspirate the buffer from around the edges of the coverslip.

(b) Carefully add 100 μl of the B cell suspension (10^5 B cells) dropwise to the top of the coverslip until it completely covers the coverslip.

(c) Cover the plate and place it in the 37 °C incubator for the desired length of time.

(d) At the end of the incubation period, tilt the plate and carefully aspirate off the liquid. Pipette 500 μl of 4% PFA against the wall of the well until it covers the entire coverslip. Fix the cells at room temperature for 10 min.

3.8 Permeabilization and Staining of B Cell-APC Conjugates on Coverslips

All of these steps can be performed either with the coverslips on parafilm or with the coverslips in the 12-well plate. Antibodies and fluorescent probes should be used at dilutions that are recommended by the supplier or that are determined empirically to provide the best ratio of signal to background fluorescence. Dilute all antibodies immediately before use. The specificity of all antibodies should be confirmed beforehand. All the steps should be performed at room temperature.

1. Using a P-200 micropipette, remove the PFA from the coverslips. Discard the PFA solution as hazardous liquid waste.

2. If staining on parafilm:

(a) Wash the coverslips three times with room temperature PBS. For each wash, gently add 200 μl of PBS to the very edge of the coverslip and make sure that it covers the entire coverslip. Aspirate off the PBS using a P-200 tip attached to a vacuum line, touching only the very edge of the coverslip.

(b) Permeabilize the cells using 100 μl of 0.2% Triton X-100 for 2 min.

(c) Remove the Triton X-100 solution and wash the coverslips three times with PBS.

(d) Cover the coverslips with 100 μl of blocking buffer (PBS with 2% BSA and 0.1% Triton X-100; *see* Subheading 2.8) for 30 min.

3. If staining in a 12-well plate:

(a) Gently pipet 500 μl of PBS against the side of the well until it covers the entire coverslip. Then tilt the plate and gently aspirate the liquid using a P-200 tip attached to a vacuum line, touching the tip to the side of the well.

(b) Permeabilize the cells using 500 μl of 0.2% Triton X-100 for 2 min.

(c) Remove the Triton X-100 solution and wash the coverslips three times with PBS.

(d) Cover the coverslips with 500 µl of blocking buffer for 30 min.

4. Dilute primary antibodies in blocking buffer to the appropriate concentrations. Primary antibodies that we have used in this procedure are listed in Subheading 2.8 and include antibodies against phosphotyrosine (P-Tyr), phospho-CD79a (P-CD79a), and α-tubulin.

5. Aspirate the blocking buffer. Remove as much liquid as possible but avoid drying out the surface of the coverslip. A thin film of blocking buffer should remain on the surface of the coverslip.

6. Add 50 µl of primary antibody solution to each coverslip. Make sure that the entire surface of the coverslip is covered. Incubate at room temperature for 30–40 min.

7. Dilute secondary antibodies in blocking buffer. Secondary antibodies that we have used are listed in Subheading 2.8.

(a) To visualize the single-chain anti-Igκ antibody (i.e., the antigen expressed on the surface of the APC), Alexa Fluor 488- or Alexa Fluor 647-conjugated goat anti-rat IgG (Fc region-specific, no cross-reactivity with mouse IgG) is added to the secondary antibody mix.

(b) If F-actin is to be visualized, fluorophore-conjugated phalloidin (*see* Subheading 2.8) is added to the secondary antibody mix.

8. Wash the coverslips three times with PBS as in **step 2a** (if staining on parafilm) or **step 3a** (if staining in a 12-well plate). After the last wash, remove most of the liquid but leave a thin film of PBS on the surface of the coverslip.

9. Add 50 µl of secondary antibody solution to each coverslip, ensuring that the entire surface is covered. Incubate at room temperature for 30–40 min with the coverslips protected from light. Cover the plate containing the coverslips with an opaque box or with aluminum foil.

10. Wash the coverslips three times with PBS as in **step 2a** or **3a**.

11. Add a small droplet (~2 mm diameter; ~10 µl) of ProLong Diamond mounting reagent (with DAPI if you also wish to visualize the nucleus) to the surface of a microscope slide.

12. Using forceps, pick up the coverslip. Gently tap the side of the coverslip against a Kimwipe to remove excess liquid.

13. Place the coverslip cell-side down on the droplet of mounting reagent. Avoid trapping bubbles between the microscope slide and the coverslip. The mounting reagent should cover the

entire coverslip. Let the slides dry overnight at room temperature, protected from light.

14. The slides can then be imaged using a fluorescent microscope or stored at 4 °C.

3.9 Imaging Fixed Cells with a Laser Scanning Confocal Microscope

1. Open the imaging software program and set the capture parameters on the confocal microscope such that the laser power and exposure times are optimized for the sample, based on the brightness of the cells. Use the lowest laser power and shortest exposure times possible to minimize photobleaching and phototoxicity. Adjust the fluorescence capture intensity and microscope gain values to maximize the dynamic range of detection while minimizing pixel saturation. All the samples for which the fluorescence intensity will be quantified and compared to each other should be imaged with identical settings.

2. Adjust the microscope focus so that the contact site between an APC and a B cell is in focus. Distinct antigen microclusters should be seen at the interface between the B cell and the APC.

3. Capture the image.

4. Optional: If you wish to obtain a *z*-stack instead of a single-confocal slice containing the B cell-APC interface, set the software parameters to capture a *z*-stack. We collect 0.3 μm *x*-*y* slices using a 100× objective lens. Select an APC with interacting B cells on the top of it and set the focus so that the contact site between the bottom of the APC and the coverslip is in focus. Set this as the lower limit of the *z*-stack. Change the focal plane so that the top of the B cell is in focus and set this as the upper limit of the *z*-stack. You can use the imaging software to generate a 3D reconstruction of the B cell interacting with the APC.

5. Use image analysis software such as ImageJ or Slidebook to quantify fluorescence signals as well as the number and size of clusters at the B cell-APC contact site (*see* Subheading 3.12).

3.10 Preparing APCs and B Cells for Live Cell Imaging

If B cells expressing fluorescent proteins are to be used, transfect them 1–2 days before imaging B cell-APC interactions. Labeling the B cells with a membrane dye should be carried out immediately before using the cells for live cell imaging. When pretreating B cells with cytoskeletal inhibitors or other drugs, the optimal pretreatment time should be determined. Short pretreatments (e.g., a few minutes) with drugs can be carried out after staining the cells with a membrane dye. However if longer pretreatments (e.g., 1 h) are required, treat the cells with the inhibitor prior to staining the B cells. If the effects of the inhibitor are rapidly reversible, ensure that the same concentration of inhibitor is present during all the subsequent steps including while staining the B cells with dyes, washing the cells, and during the B cell-APC interaction.

If using APCs expressing HEL-HaloTag as the antigen, the Cos-7 cells must be labeled with the Halo-Tag ligand prior to live cell imaging. The following procedure has been adapted from the manufacturer's instructions and optimized for imaging B cell: APC interactions.

1. Cos-7 cells that have been transfected with HEL-HaloTag and seeded onto fibronectin-coated coverslips in a 12-well tissue culture plate (Subheading 3.3) should have spread and flattened. Transfer these coverslips onto a piece of parafilm.

2. Add 190 μl of mHBS-FCS to the center of the coverslip.

3. Dilute the HaloTag ligand 1:1000 in mHBS-2% FCS. Add 5 μl of this diluted HaloTag ligand to the mHBS-FCS on the coverslip.

4. Place the coverslips in a 37 °C tissue culture incubator for 15 min.

5. Wash the coverslips three times with mHBS-2% FCS.

6. The HEL-HaloTag-expressing Cos-7 APCs are now ready to be used for live cell imaging.

1. Count the B cells and centrifuge 10^6 cells at 1500 rpm ($525 \times g$) for 5 min at room temperature.

2. Aspirate the supernatant and resuspend the cells in 1 ml of PBS.

3. Stain the cells with CellMask Far Red membrane dye (1:1000 final dilution). Alternatively, the cells can be stained with CELLVUE Maroon membrane dye according to the manufacturer's instructions.

4. Incubate the cells at 37 °C for 10 min protected from light.

5. Add 5 ml of PBS-FCS (PBS with 2% FCS) and pipette the cells up and down to mix well. Centrifuge the cells for 5 min at 1500 rpm ($525 \times g$) at room temperature. Aspirate the supernatant.

6. Wash the cells twice more with 5 ml of PBS-FCS, as in **step 5**, to remove excess dye.

7. Aspirate the final PBS-FCS wash and resuspend the cells in mHBS-FCS so that the final cell concentration is ~10^6 cells/ml. Note: Up to 50% of cells can be lost during the wash steps.

8. Leave the cells on ice, protected from light, for up to 2 h until you are ready to perform live cell imaging. Keeping the cells on ice ensures that the membrane dye is not internalized.

9. Warm the cells for 5 min in a 37 °C water bath before adding them to the APCs and imaging.

3.10.3 Treating B Cells with Cytoskeletal Inhibitors

The optimal concentrations of cytoskeletal inhibitors should be determined empirically. We provide some recommendations below that are optimal for primary B cells and A20/D1.3 cells.

1. Just before adding the B cells to the APCs, add the cytoskeletal inhibitor to the B cells and incubate for 5 min at 37 °C.

    (a) The microtubule-depolymerizing drug nocodazole is typically used at a final concentration of 1–5 μM. The actin-depolymerizing drug latrunculin A is typically used at a final concentration of 1 μM.

    (b) The effects of these drugs are reversible. Hence, it is important to maintain the same concentration of the drug during the B cell-APC interactions. The possibility that the drugs also affect the APCs should be taken into account.

3.11 B Cell-APC Interactions: Real-Time Imaging of Antigen (HEL-GFP or HEL-HaloTag) Cluster Formation Using Spinning Disk Microscopy

Begin heating the microscope imaging incubator chamber to 37 °C early in the day (at least 1 h in advance) so that the chamber temperature is properly equilibrated. The immersion oil should also be warmed to 37 °C.

1. Disassemble a Chamlide magnetic imaging chamber.

2. Using forceps, remove a coverslip with adhered Cos-7 cells from the 12-well plate and place the coverslip onto the bottom plate of the magnetic imaging chamber with the cells facing up.

3. Place the silicon O-ring into the underside of the magnetic imaging chamber main body. Then bring the main body of the imaging chamber close to the bottom plate such that the magnetic force causes it to gently snap into place. Take care not to allow the main body of the magnetic chamber to snap in place with too much force as this could break the coverslip.

4. Add 150 μl of mHBS-FCS (pre-warmed to 37 °C) to the center of the coverslip.

5. Apply a small droplet of immersion oil to the 100× objective of the microscope.

6. Wet a Kimwipe with 70% ethanol and use it to wipe the bottom of the coverslip.

7. Gently place the Chamlide magnetic imaging chamber into the universal mounting frame on the microscope stage. Adjust the clamps on the mounting frame to hold the chamber tightly in place.

8. Using the fluorescence signal from the HEL-GFP or HEL-HaloTag antigen, find a field of view with a Cos-7 cell that is flat and has a relatively uniform distribution of antigen (Fig. 3).

9. Set the capture parameters on the spinning disk confocal microscope such that the laser power and exposure times are optimized for the sample. Use the lowest laser power and shortest exposure times possible in order to minimize photobleaching

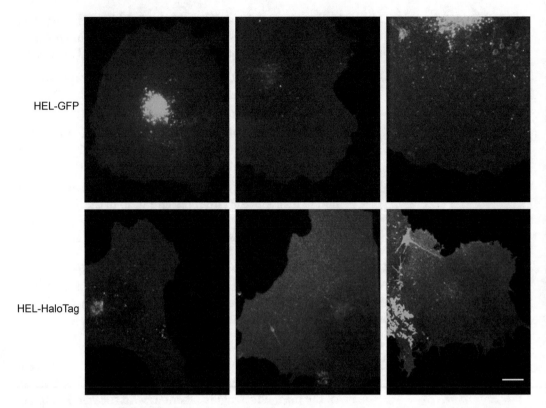

HEL-GFP

HEL-HaloTag

Fig. 3 Expression of model antigens on Cos-7 cells. Three examples are shown of Cos-7 APCs expressing the model antigens HEL-GFP (*top*) or HEL-HaloTag (*bottom*). The cells were allowed to spread on fibronectin-coated coverslips and were imaged with a 100× objective. The antigen is uniformly distributed across the surface of the Cos-7 cells, except in the middle of the cell, where the fluorescent antigen is highly concentrated in the Golgi apparatus and associated intracellular vesicles. B cell-APC interactions should be imaged in the regions of the Cos-7 cells where the antigen distribution is relatively uniform. Scale bar: 10 μm

and phototoxicity. Adjust the fluorescence capture intensity and microscope gain values to maximize the dynamic range of detection while minimizing pixel saturation. All the samples for which the fluorescence intensity will be quantified and compared to each other should be imaged with identical settings.

10. If using cytoskeletal inhibitors or other drugs, add the drug (or solvent control) to the 150 μl of buffer in the imaging chamber immediately before adding the B cells to the APCs. The final concentration of the drug in the 150 μl of buffer should be the same as what has been added to the B cells (*see* Subheading 3.10.3).

11. Add 100 μl of the B cell suspension (10^5 cells) dropwise to the center of the imaging chamber, being careful not to disturb the Cos-7 APCs or to shift the field of view. Begin imaging immediately.

12. Within 1–3 min, control B cells should settle onto the APCs and begin forming antigen clusters at the contact site (Fig. 4).

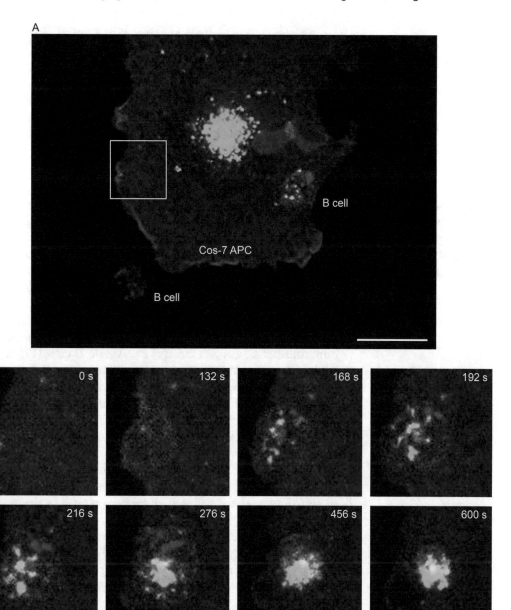

Fig. 4 Live cell imaging of antigen cluster formation at the B cell-APC interface. B cells (*red*) were allowed to settle onto Cos-7 APCs expressing the HEL-GFP (*green*) model antigen. Imaging was commenced immediately after adding the B cells to the APCs. (**a**) Representative image showing an APC to which a B cell has bound and started to form microclusters. Another B cell has settled on the coverslip and not on the APC. In subsequent frames a B cell bound to the region of the APC that is enclosed by the *white box*. Scale bar: 10 μm. (**b**) Time-lapse images of the region of the APC enclosed by the *white box* in (a) showing a B cell binding to the APC, the formation of BCR/antigen microclusters and their subsequent coalescence into a cSMAC. Time points (in seconds) are shown on each panel. Scale bar: 3 μm

These clusters will then coalesce into a large cluster at the center of the contact site to form the cSMAC. *See* Wang et al. [16] for movies showing control and nocodazole-treated B cells forming antigen clusters on APCs.

13. If a B cell does not contact the APC that is in the field of view within 5 min, disassemble the Chamlide imaging chamber, discard the coverslip, and begin again from **step 2** until a B cell-APC interaction is observed.

14. Export image files as 16-bit TIFF images for image analysis.

3.12 Quantifying Antigen Clusters and BCR Signaling at the B Cell-APC Contact Site

Both ImageJ plug-ins and custom MATLAB scripts can be used for quantitative analysis of BCR organization and signaling at the B cell-APC contact site. For each B cell that has contacted an APC, the parameters that can be quantified include (1) the total amount of antigen (in arbitrary fluorescence units) that has been gathered into clusters at the B cell-APC contact site, (2) the total amount of proximal BCR signaling (e.g., phosphorylation of the CD79a/CD79b [Ig-α/Ig-β] subunit of the BCR or phosphotyrosine [P-Tyr]) that is present as clusters at the B cell-APC contact site, (3) the number of antigen clusters at a single B cell-APC contact site as well as their individual sizes and fluorescence intensity, and (4) the number, size, and fluorescence intensities of P-CD79a or P-Tyr clusters at the B cell-APC contact site. This analysis can be extended to other proteins or signaling reactions that can be detected at the B cell-APC contact site by immunofluorescence or by using fluorescent probes.

The following image analysis method employs ImageJ to determine for individual B cells the total amount of gathered antigen and BCR signaling at the B cell-APC interface, as well as the size and number of clusters. This analysis requires that the contact site between the B cell and the APC is in a single *x-y* focal plane. In addition to analyzing images from fixed cells, this method can be applied to still images from live cell imaging videos. To do this, individual frames from the video can be imported into ImageJ.

1. Use the microscope's image acquisition software to export the image files as 16-bit TIFF files.

2. Import the TIFF images for the antigen channel, and for any other fluorescent signal that has been imaged (e.g., the P-Tyr or P-CD79a channel), into ImageJ (Fig. 5a). Carry out **steps 3–10** to determine the total amount of antigen fluorescence that has been gathered into clusters at the B cell-APC interface. Repeat **steps 3–10** for the other channels (e.g., P-Tyr or P-CD79a) to determine the total amount of fluorescence that is present as clusters at the B cell-APC interface.

3. Adjust the image threshold so that only the clusters of antigen, P-Tyr, or P-CD79a are overlaid with a red mask while the

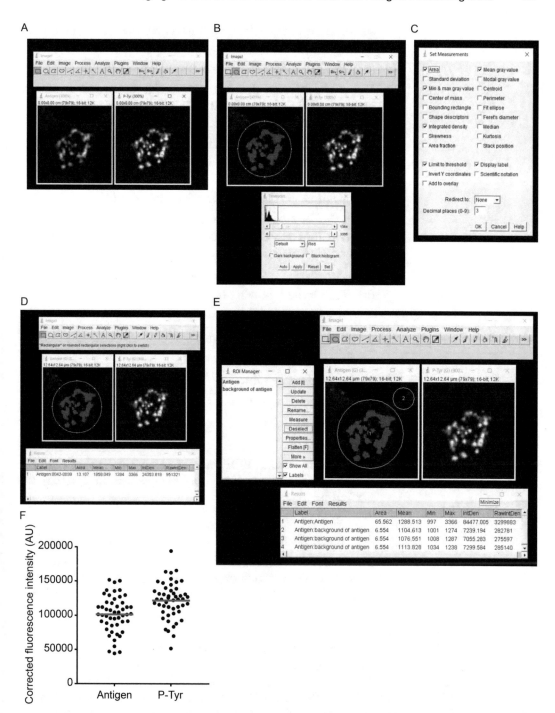

Fig. 5 Using ImageJ to quantify the fluorescence intensity of antigen clusters and phosphotyrosine staining at the B cell-APC contact site. (**a**) 16-bit TIFF images of antigen (HEL-GFP, *left*) and P-Tyr (*right*) that have been imported into ImageJ. (**b**) The threshold was adjusted to create a red mask that demarks the regions containing the antigen clusters. A circular ROI was drawn around the antigen clusters associated with a single B cell. (**c**) The Set Measurements window in ImageJ was used to choose the parameters to be measured. (**d**) The total antigen fluorescence in the red-masked areas with the ROI was determined and now

surrounding regions lacking clusters are black (i.e., below the detection threshold) (Fig. 5b).

   (a) *Image → Adjust → Threshold*

4. Using the "Oval" drawing tool, draw a circular region of interest (ROI) around the B cell, being sure to encompass all of the clusters that are overlaid with the red mask (Fig. 5b).

5. Open the *ROI Manager* window and add the circular ROI to the manager in order to save it and apply it to other images that you will analyze. To do this, use the following commands:

   (a) *Analyze → Tools → ROI Manager → Add*

6. Open the *Set Measurements* dialogue box, and set the measurements so that only the sections within the ROI that are overlaid with a red mask are quantified (Fig. 5c).

   (a) *Analyze → Set Measurements*

   (b) Check the following boxes:

      i. Area

      ii. Min and max gray value

      iii. Integrated density

      iv. Mean gray value

      v. Limit to threshold

      vi. Display label

   (c) Select *OK*.

7. Click on the circular ROI drawn in **step 4** to ensure that it is selected. Measure the total fluorescence intensity within the masked regions.

   (a) *Analyze → Measure*

   (b) The measurements will appear in a new window (Fig. 5d).

8. Determine the background fluorescence value that will be subtracted from the total fluorescence value that was determined in **step** 7. The background correction will use the average

Fig. 5 (continued) appears in the window "Results" below the image. (**e**) The "ROI Manager" in ImageJ (*upper left*) was used to save the antigen and background ROIs. The results window (*bottom*) displays the fluorescence intensities for the antigen clusters and the three background ROIs. The average value for the three background ROIs is subtracted from the antigen fluorescence intensity to generate the corrected fluorescence intensity for the antigen. The same steps depicted in panels **b–e** are used to determine the corrected fluorescence intensity for the P-Tyr signal. (**f**) Graphs of simulated data showing the background-corrected fluorescence intensities for the antigen clusters and P-Tyr signal. Each dot is the value for an individual B cell and the red lines represent the mean values. The values shown in these graphs were generated randomly to reflect a Gaussian distribution (https:/www.random.org/gaussian-distributions/). The values generated from actual data may not have a normal distribution. Also, because outliers unduly influence the mean value, it may be more appropriate to report median values

fluorescence intensity for three regions of the APC where there are no B cells and no fluorescent clusters in that channel. In the *Set Measurements* dialogue box (Fig. 5c), uncheck "Limit to threshold." Draw an ROI of any size at a nearby site on the APC where there are no clusters (Fig. 5e). Note that the area of the background ROI will not be considered for correcting the background fluorescence, only the mean gray value for the background ROI.

(a) *Analyze → Set Measurements →* uncheck "Limit to threshold."

(b) Select the background ROI using the cursor.

(c) *Analyze → Measure.*

9. Repeat **step 8** at least twice more, moving this background ROI to other nearby locations on the APC that do not overlap the red masked area. At least three background measurements should be acquired (Fig. 5e).

10. In Excel, subtract the background fluorescence from the total fluorescence using the following equation, as described in [16] to yield the total background-corrected fluorescence intensity:

Total background-corrected fluorescence intensity $= (A_{\text{clusters}})$ $(m_{\text{clusters}} - \bar{m}_{\text{background}})$
where:

A_{clusters} is the area of the red-masked area that is overlaid on the clusters,

m_{clusters} is the mean fluorescence intensity per pixel in the red-masked areas, and

$\bar{m}_{\text{background}}$ is the average value for the mean fluorescence intensities for the background ROIs.

11. The data can be presented as dot plots showing the corrected fluorescence intensity, in arbitrary units (AU), for each B cell that was analyzed using **steps 3–10** (Fig. 5f).

12. To determine the size, number, and fluorescence intensity of individual antigen, P-Tyr, or P-CD79a clusters at a single B cell-APC contact site, click on the circular ROI drawn in **step 4** to select it and proceed with the following commands:

(a) *Analyze → Analyze Particles*

A new window will open. Set the following parameters:

i. Size (μm^2): 0-Infinity

ii. Circularity: 0.00–1.00

iii. "Display results" checked

iv. "Summarize" checked

v. "Add to manager" checked

(b) *Click OK.*

13. Two new windows will open:

  (a) *Summary* window: These data will indicate how many clusters were counted and their average size.

  (b) *Results* window: This window will display the size and fluorescence intensity values for individual clusters.

Acknowledgments

This work was supported by grant MOP-68865 (to M.R.G) from the Canadian Institutes of Health Research (CIHR). J.C.W. was supported by a doctoral scholarship from the Natural Sciences and Engineering Research Council of Canada (NSERC). M. B.-M. was supported by a doctoral scholarship from the University of British Columbia.

References

1. Batista FD, Harwood NE (2009) The who, how and where of antigen presentation to B cells. Nat Rev Immunol 9:15–27

2. Gonzalez SF, Degn SE, Pitcher LA et al (2011) Trafficking of B cell antigen in lymph nodes. Annu Rev Immunol 29:215–233

3. Cyster JG (2010) B cell follicles and antigen encounters of the third kind. Nat Immunol 11:989–996

4. Wykes M, Pombo A, Jenkins C et al (1998) Dendritic cells interact directly with naive B lymphocytes to transfer antigen and initiate class switching in a primary T-dependent response. J Immunol 161:1313–1319

5. Qi H, Egen JG, Huang AY et al (2006) Extra-follicular activation of lymph node B cells by antigen-bearing dendritic cells. Science 312:1672–1676

6. Colino J, Shen Y, Snapper CM (2002) Dendritic cells pulsed with intact Streptococcus pneumoniae elicit both protein- and polysaccharide-specific immunoglobulin isotype responses in vivo through distinct mechanisms. J Exp Med 195:1–13

7. Batista FD, Iber D, Neuberger MS (2001) B cells acquire antigen from target cells after synapse formation. Nature 411:489–494

8. Fleire SJ, Goldman JP, Carrasco YR et al (2006) B cell ligand discrimination through a spreading and contraction response. Science 312:738–741

9. Harwood NE, Batista FD (2010) Early events in B cell activation. Annu Rev Immunol 28:185–210

10. Carrasco YR, Fleire SJ, Cameron T et al (2004) LFA-1/ICAM-1 interaction lowers the threshold of B cell activation by facilitating B cell adhesion and synapse formation. Immunity 20:589–599

11. Carrasco YR, Batista FD (2006) B-cell activation by membrane-bound antigens is facilitated by the interaction of VLA-4 with VCAM-1. EMBO J 25:889–899

12. Treanor B, Depoil D, Gonzalez-Granja A et al (2010) The membrane skeleton controls diffusion dynamics and signaling through the B cell receptor. Immunity 32:187–199

13. Harwood NE, Batista FD (2011) The cytoskeleton coordinates the early events of B-cell activation. Cold Spring Harb Perspect Biol 3: a002360

14. Treanor B, Depoil D, Bruckbauer A et al (2011) Dynamic cortical actin remodeling by ERM proteins controls BCR microcluster organization and integrity. J Exp Med 208:1055–1068

15. Freeman SA, Lei V, Dang-Lawson M et al (2011) Cofilin-mediated F-actin severing is regulated by the Rap GTPase and controls the cytoskeletal dynamics that drive lymphocyte spreading and BCR microcluster formation. J Immunol 187:5887–5900

16. Wang JC, Lee JY, Christian S et al (2017) The Rap1-cofilin-1 pathway coordinates actin reorganization and MTOC polarization at the B-cell immune synapse. J Cell Sci 130:1094–1109

17. Schnyder T, Castello A, Feest C et al (2011) B cell receptor-mediated antigen gathering requires ubiquitin ligase Cbl and adaptors

Grb2 and Dok-3 to recruit dynein to the signaling microcluster. Immunity 34:905–918

18. Yuseff M-I, Reversat A, Lankar D et al (2011) Polarized secretion of lysosomes at the B cell synapse couples antigen extraction to processing and presentation. Immunity 35:361–374

19. Ketchum C, Miller H, Song W et al (2014) Ligand mobility regulates B cell receptor clustering and signaling activation. Biophys J 106:26–36

20. Lin KB, Freeman SA, Zabetian S et al (2008) The Rap GTPases regulate B cell morphology, immune-synapse formation, and signaling by particulate B cell receptor ligands. Immunity 28:75–87

21. Dustin ML (2009) Supported bilayers at the vanguard of immune cell activation studies. J Struct Biol 168:152–160

22. Comrie WA, Babich A, Burkhardt JK (2015) F-actin flow drives affinity maturation and spatial organization of LFA-1 at the immunological synapse. J Cell Biol 208:475–491

23. Tolar P, Spillane KM (2014) Force generation in B-cell synapses: mechanisms coupling B-cell receptor binding to antigen internalization and affinity discrimination. Adv Immunol 123:69–100

24. Natkanski E, Lee W-Y, Mistry B et al (2013) B cells use mechanical energy to discriminate antigen affinities. Science 340:1587–1590

25. Nowosad CR, Tolar P (2017) Plasma membrane sheets for studies of B cell antigen internalization from immune synapses. Methods Mol Biol 1584:77–88

26. Usui K, Honda S, Yoshizawa Y et al (2012) Isolation and characterization of naive follicular dendritic cells. Mol Immunol 50:172–176

27. Ait-Azzouzene D, Verkoczy L, Peters J et al (2005) An immunoglobulin C kappa-reactive single chain antibody fusion protein induces tolerance through receptor editing in a normal polyclonal immune system. J Exp Med 201:817–828

28. Freeman SA, Jaumouille V, Choi K et al (2015) Toll-like receptor ligands sensitize B-cell receptor signalling by reducing actin-dependent spatial confinement of the receptor. Nat Commun 6:6168

29. Goodnow CC, Crosbie J, Adelstein S et al (1988) Altered immunoglobulin expression and functional silencing of self-reactive B lymphocytes in transgenic mice. Nature 334:676–682

30. Batista FD, Neuberger MS (1998) Affinity dependence of the B cell response to antigen: a threshold, a ceiling, and the importance of off-rate. Immunity 8:751–759

31. Yi J, Wu X, Chung AH et al (2013) Centrosome repositioning in T cells is biphasic and driven by microtubule end-on capture-shrinkage. J Cell Biol 202:779–792

Chapter 11

In Vivo Tracking of Particulate Antigen Localization and Recognition by B Lymphocytes at Lymph Nodes

Yolanda R. Carrasco

Abstract

The development of experimental systems that allow in vivo antigen tracking as well as the study of B cell dynamics in real time and in situ, have transformed our understanding of the "how, when and where" B lymphocytes find antigen at secondary lymphoid organs in the last 10 years. Here, I described one of these experimental models, which uses highly fluorescent particulate antigen and B cell receptor (BCR)-transgenic B cells labeled with long-term fluorescent probes, combined with confocal and multiphoton microscopy.

Key words Particulate antigen, B cells, In vivo tracking

1 Introduction

Since the application of multiphoton (MP) microscopy to the study of the immune system in 2002 [1], scientists have developed distinct experimental models and strategies for the visualization of B lymphocyte biology in real time and in situ. In the last 15 years our understanding of how, where, and when naive B cells encounter antigen in vivo greatly increased [2–5], as well as how B cells seek for T cell help [6] and the germinal center reaction takes place [7, 8]. All that information revolutionized the established paradigms and compelled scientists to revisit them.

Studies on the initial stages of B cell antigen recognition using in vitro models pointed out the B cell ability to recognize native antigen presented on the surface of antigen-presenting cells, to establish the Immune Synapse, and the relevance of this platform for B cell function [9–11]. The necessity to validate the in vitro observations in the "real life" led us and other laboratories to develop experimental strategies to visualize and track in vivo the process of B cell antigen encounter in lymph node (LN) follicles [2–4].

Chaohong Liu (ed.), *B Cell Receptor Signaling: Methods and Protocols*, Methods in Molecular Biology, vol. 1707, https://doi.org/10.1007/978-1-4939-7474-0_11, © Springer Science+Business Media, LLC 2018

Here, I explained in detail the experimental setup we developed to monitor B cell recognition of particulate antigen in vivo at LN [2]. We used fluorescent spheres of the size of a virus coated with the antigen of interest as particulate antigen to immunize mice. We isolated primary B cells from wild type (WT) and BCR transgenic mouse models, labeled them with long-term fluorescent probes, and adoptively transferred them to recipient mice before particulate antigen administration. At distinct time points after particulate antigen inoculation at the mouse footpad, we isolated draining LN and analyzed them either by tissue-sections immunofluorescence and confocal microscopy or by MP microscopy of the whole explanted organ in real time. We analyzed the acquired data using cell tracking and imaging analysis software.

2 Materials

Find below the materials, reagents, and buffers required.

2.1 Preparation of Fluorescently Labeled Primary B Cells

1. Lympholyte cell separation media for mice (Cedarlane Lab).
2. Kit for B cell negative selection; I used the one from Dynal, but other companies also supply them (Miltenyi, StemCell).
3. Erythrocyte Lysis Buffer.
4. Cell-Tracker probes for long-term tracing of living cells (SNARF-1, CMAC; Molecular Probes). Prepare stock at 1–10 mM in DMSO, aliquot and store at −20 °C (see **Note 1**).
5. PBS pH 7.4 (GIBCO), Fetal Calf Serum (FCS, complement-inactivated), and RPMI medium.

2.2 Preparation of Particulate Antigen

1. FluoSpheres Neutravidin-labeled Microspheres, 0.2 μm, yellow–green fluorescent (505/515), 1% solids (F8774, Molecular Probes) (see **Note 2**).
2. Hen Egg Lysozime (HEL, Sigma), prepare a stock of 1 mg/ml in PBS and keep it at 4 °C.
3. NHS-Biotin-LC-LC (Amersham), stock at 200 mg/ml in DMSO. Aliquot and keep it at −20 °C.
4. Wash buffer: PBS pH 7.4 with 1% BSA (Sigma) 0.02% NaN$_3$. Once prepared, filter it in the tissue culture hood using a MCE filter of 0.22 μm to obtain a sterile buffer. Keep it at 4 °C.

2.3 Immunofluorescence (IF) of Tissue Sections from the Draining Lymph Node

1. Liquid nitrogen.
2. 4% Paraformaldehide (PFA) in PBS, pH 7–7.5, freshly prepared and filtered.
3. IF-Blocking buffer: PBS 1% BSA 10% goat serum.
4. Fluoromount (Southern Biotech).

2.4 Preparation of Lymph Nodes for Multiphoton (MP) Microscopy in Real Time

1. Glass-bottom 35 mm culture dish (Mattek).
2. Veterinary topical tissue adhesive (Nexaband, WPI).
3. MP-medium: RPMI without Phenol Red (GIBCO).
4. A bottle of Carbogen (95% O_2, 5% CO_2).

3 Methods

3.1 Adoptive Transfer of Primary B Cells into Recipient Mice

Work in a flow hood for sterile conditions.

1. Remove the spleens from 3–5 months-old WT and BCR-transgenic MD4 mice (BCR specific for HEL; [12]), both in C57BL/6 genetic background. Homogenize them separately in PBS (10 ml), pass the cell suspension through a cell strainer (70 μm), and centrifuge over a layer of Lympholyte (2 ml; 1,769 × g, 20 min, room temperature (RT), without brake) using 15 ml tubes.

2. Collect the interphase, highly enriched in viable lymphocytes, wash in RPMI 10% FCS (546 × g, 8 min, RT), eliminate residual erythrocytes using Erythrocyte Lysis Buffer (1 ml/spleen, 5 min, RT), wash again in RPMI 10% FCS (349 × g, 5 min, RT) and follow kit instructions to purified B cells.

3. Label purified B cells (5 × 10^6/ml) with 2 μm SNARF-1 or 50 μm CMAC in PBS (15 min, 37 °C, in 15 ml tubes), stop labeling reaction by adding one volume of FCS (1 min, RT), wash in RPMI 10% FCS (349 × g, 8 min, RT), and leave in culture in RPMI 10% FCS for 1 h for dye equilibration.

4. Adoptively transfer 5–10 millions of SNARF-1-labeled MD4 B cells or 5 millions of SNARF-1-labeled MD4 B cells plus 5 millions of CMAC-labeled WT B cells (0.2 ml PBS) by tail-vein injection into 4–6 week old C57BL/6 recipient mice. Leave them 20 h before particulate antigen inoculation.

3.2 Inoculation of Particulate Antigen into the Footpad of Recipient Mice

1. Prepare in advance the biotin-labeled antigen, in this case HEL, as follows. Incubate 1 ml of HEL (1 mg/ml, in PBS) with NHS-biotin reagent at 1 μg/ml final concentration (30 min, RT, in darkness). Then, dialyze against PBS (1 l), exchanging 4 times the PBS volume. Keep the biotin-labeled HEL at 4 °C.

   Use a flow hood to work in sterile conditions. All buffers need to be sterile. Prepare the particulate antigen suspension the same day of inoculation.

2. Vortex the vial of fluospheres. Add 1 μl fluospheres to 1 ml wash buffer, in eppendorf tubes, and centrifuge (20,800 × g, 10 min, RT; you should see the pellet) (*see* **Notes 3** and **4**).

3. Incubate with biotin-labeled HEL (1 μg) in wash buffer (0.2 ml), shaking (20 min, RT).

4. Wash twice in wash buffer (1 ml; 20,800 × g, 10 min, RT). Do a third wash in PBS (20,800 × g, 10 min, RT; you should see the pellet).

5. Resuspend in 25 μl PBS for injection into mouse footpad. The prepared particulate antigen can be kept at 4 °C until injection. Warm it up before injection.

6. Inoculate the mice (previously transferred with B cells). Anesthetize mouse, and inoculate the particulate antigen suspension (25 μl; shake it before injection) using insulin needles or 25G 5/8" needles (*see* **Note 5**).

3.3 Preparation of Draining (Popliteal) Lymph Node for Immunofluorescence

1. Sacrifice mice by cervical dislocation at the selected time points. Remove popliteal LN and, if required, eliminate the surrounding fat-tissue (*see* **Notes 6** and **7**).

2. Immerse the LN in OCT medium and freeze in liquid nitrogen.

3. Make 10 μm thickness sections of the LN using a cryostat; keep sections at −80 °C.

4. Remove sections from −80 °C and allow them to come to RT in a sealed box containing dessicant (silica beads; 5–10 min).

5. Fix them in 4% PFA (10 min, RT); wash in PBS, block in IF-blocking buffer (30 min, RT), incubate primary antibody in PBS 1% BSA (1 h, RT), wash in PBS (3 times, 3 min each), incubate secondary antibody (30 min, RT), wash in PBS (3 times, 3 min each), and mount them using Fluoromount.

6. Keep the slides at 4 °C. Image using a fluorescent confocal microscope.

3.4 Preparation of Draining (Popliteal) Lymph Node for Multiphoton Microscopy

Image the mice one by one (two popliteal LN each). Always use warmed MP-medium.

1. Sacrifice and remove popliteal LN as in Subheading 3.3, and without damaging the organ eliminate the surrounding fat tissue. While imaging one LN, the other one from the same mouse can be kept at 37 °C in MP-medium (*see* **Notes 6** and **7**).

2. Attach the lymph node through the hylum to the base of the glass-bottom dish using a small drop of Nexaband; immediately after, cover the organ with warmed MP-medium, and go to the multiphoton microscope.

3. Keep the LN continuously at 37 °C and perfused with warmed MP-medium bubbled with carbogen while imaging, pumping

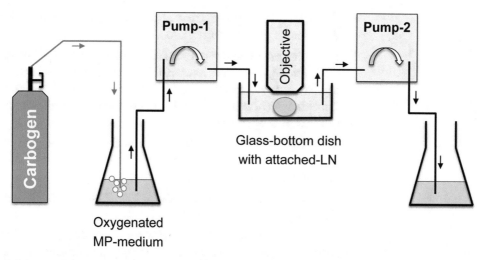

Fig. 1 Scheme of the experimental setup for MP microscopy of explanted LN. For the imaging of explanted LN by MP microscopy, you need to assemble a circuit similar to the one shown in the scheme. One recipient containing MP-medium will be bubbled with carbogen and connected to a first (peristaltic) pump to inject MP-medium into the LN-containing dish. A second (peristaltic) pump will remove MP-medium from the dish and transfer it to a second recipient. Both recipients and LN-containing dish should be inside the environmental chamber of the microscope to keep the temperature at 37 °C. The MP-medium collected in the second recipient can be reused during the experiment

in and out the MP-medium to keep an appropriate level of oxygen saturation (see Fig. 1) (*see* **Note 8**).

4. Acquire time series (20–30 times) of a z-stack of 50 μm (5 μm optical slice), at 512×512 pixels. Perform imaging through the LN capsule using an upright multiphoton microscope and a multi-immersion $25 \times /0.75$ objective. Maximum imaging time per LN: 90 min.

4   Notes

1. Avoid more than two thaw-freezing cycles of the Cell-tracker probes; they lose labeling activity.

2. The high fluorescence intensity and stability of the spheres allowed an easy and sharp detection of the particulate antigen by both types of microscopy, confocal and MP. Using confocal, you can perform quantitative studies (count n° fluospheres acquired per B cell) at early time points after particulate antigen inoculation. The particle brightness also facilitates detection at LN deeper zones by MP.

3. Fluospheres tend to form clumps; you can sonicate them in a water bath before use.

4. You can scale the preparation of particulate antigen up to 5 µl of fluospheres per eppendorf tube.

5. I used to inoculate the particulate antigen in the right footpad, for example, and fluospheres not loaded with antigen (HEL) in the left footpad from the same mouse as control. The control fluospheres (non-antigen) will be prepared in parallel with the ones to be loaded with antigen.

6. The main draining LN to analyze is the popliteal, but some analysis can also be done in the following draining LN, the inguinal, as some particulate antigen arrives.

7. Particulate antigen can be tracked at least up to 48 h after inoculation.

8. Both MP-medium bubbling with carbogen (i.e., oxygen saturation of the MP-medium) and its continuous exchange through the pumping are critical for cell motility. Slow rates of carbogen bubbling and of pumping should be fine, but I recommend you to test both when you set up the system for first time.

Acknowledgments

This work was supported by grants from the Spanish Ministry of Economy (BFU2013-48828-P) and from the Worldwide Cancer Research (WCR; grant reference number 15-1322).

References

1. Miller MJ, Wei SH, Parker I, Cahalan MD (2002) Two-photon imaging of lymphocyte motility and antigen response in intact lymph node. Science 296(5574):1869–1873

2. Carrasco YR, Batista FD (2007) B cells acquire particulate antigen in a macrophage-rich area at the boundary between the follicle and the subcapsular sinus of the lymph node. Immunity 27(1):160–171

3. Phan TG, Grigorova I, Okada T, Cyster JG (2007) Subcapsular encounter and complement-dependent transport of immune complexes by lymph node B cells. Nat Immunol 8(9):992–1000

4. Junt T, Moseman EA, Iannacone M, Massberg S, Lang PA, Boes M, Fink K, Henrickson SE, Shayakhmetov DM, di Paolo NC, van Rooijen N, Mempel TR, Whelan SP, von Andrian UH (2007) Subcapsular sinus macrophages in lymph nodes clear lymph-borne viruses and present them to antiviral B cells. Nature 450(7166):110–114

5. Roozendaal R, Mempel TR, Pitcher LA, Gonzalez SF, Verschoor A, Mebius RE, von Andrian UH, Carroll MC (2009) Conduits mediate transport of low-molecular-weight antigen to lymph node follicles. Immunity 30(2):264–276. https://doi.org/10.1016/j.Immuni.2008.12.014

6. Okada T, Miller MJ, Parker I, Krummel MF, Neighbors M, Hartley SB, O'Garra A, Cahalan MD, Cyster JG (2005) Antigen-engaged B cells undergo chemotaxis toward the T zone and form motile conjugates with helper T cells. PLoS Biol 3(6):e150

7. Allen CD, Okada T, Tang HL, Cyster JG (2007) Imaging of germinal center selection events during affinity maturation. Science 315(5811):528–531

8. Schwickert TA, Lindquist RL, Shakhar G, Livshits G, Skokos D, Kosco-Vilbois MH, Dustin ML, Nussenzweig MC (2007) In vivo imaging of germinal centres reveals a dynamic open structure. Nature 446(7131):83–87

9. Batista FD, Iber D, Neuberger MS (2001) B cells acquire antigen from target cells after synapse formation. Nature 411(6836):489–494

10. Carrasco YR, Fleire SJ, Cameron T, Dustin ML, Batista FD (2004) LFA-1/ICAM-1 interaction lowers the threshold of B cell activation by facilitating B cell adhesion and synapse formation. Immunity 20(5):589–599

11. Fleire SJ, Goldman JP, Carrasco YR, Weber M, Bray D, Batista FD (2006) B cell ligand discrimination through a spreading and contraction response. Science 312 (5774):738–741. https://doi.org/10.1126/science.1123940

12. Goodnow CC, Crosbie J, Adelstein S, Lavoie TB, Smith-Gill SJ, Brink RA, Pritchard-Briscoe H, Wotherspoon JS, Loblay RH, Raphael K et al (1988) Altered immunoglobulin expression and functional silencing of self-reactive B lymphocytes in transgenic mice. Nature 334(6184):676–682

Chapter 12

Study B Cell Antigen Receptor Nano-Scale Organization by In Situ Fab Proximity Ligation Assay

Kathrin Kläsener, Jianying Yang, and Michael Reth

Abstract

The B cell antigen receptor (BCR) is found to be non-randomly organized at nano-scale distances on the B cell surface. Studying the organization and relocalization of the BCR is thus likely to provide new clues to understand the activation of the BCR. Indeed, with the in situ Fab proximity ligation assay (Fab-PLA), we now obtain proofs for the dissociation activation of BCRs and start to gain insight into how the relocalization of B cell surface signaling molecules could activate the cells. This chapter describes our methods to study the nano-scale organization of B cell surface receptors and co-receptors with Fab-PLA.

Key words BCR, Nano-scale organization, Fab-PLA

1 Introduction

For long time, it was thought that the BCR is activated by the crosslinking of two BCR monomers through their binding to multivalent stimuli such as polyvalent antigens and anti-BCR antibodies [1]. However, in recent years, it was found that many membrane proteins including the BCR have more complex oligomeric structures and that they are non-randomly organized at nano-scale distances [2–5]. Unfortunately, the techniques allowing the study of the nano-scale membrane protein organization and interaction are still limited. Recent development of super-resolution microscopy enabled researchers to obtain images with sub-hundred-nanometer (nm) resolution [6]. As super-resolution techniques are limited by labeling efficiency and overcounting problems [7], it is still impossible to get a bona fide super-resolution snapshot for the oligomeric organization of membrane receptors. In fact, several recent super resolution microscopic studies of BCRs in the nano-scale have produced data that conflict with each other [8–11], probably due to these unsolved problems.

The in situ proximity ligation assay (PLA) detects the close, nano-scale proximity of two target proteins by amplifying the signal

Chaohong Liu (ed.), *B Cell Receptor Signaling: Methods and Protocols*, Methods in Molecular Biology, vol. 1707, https://doi.org/10.1007/978-1-4939-7474-0_12, © Springer Science+Business Media, LLC 2018

through PCR after ligating oligos coupled to antibodies indirectly (2-PLA) or directly (1-PLA) [12]. The usage of antibody to detect the target protein in this assay promises the identification of the close proximity of two targets without the need of genetically engineered proteins, hereby PLA allows studying protein–protein interaction under their native forms, such as ex vivo samples. However, considering the size of an antibody, it is worth noting that the maximum detection range for 2-PLA and 1-PLA is 80 to 40 nm and thus much bigger than the size of a BCR complex (~25 nm). However, by using Fab-fragments instead of whole antibodies, the detection range of PLA can be improved. The Fab-PLA technique has been successfully employed to monitor the surface organization and relocalization of BCRs and co-receptors [13, 14]. The results from these studies have provided direct proofs for the opening of BCR oligomers and the nanoscale reorganization of surface signaling molecules including co-receptors during BCR activation. In this chapter, we describe a procedure for studying the organization and relocalization of BCR and other signaling molecules using in situ Fab-PLA.

2 Materials

1. Antibodies: Fab-PLA is successfully used to quantify the oligomerization level of IgM- and IgD-BCR, the interaction between IgM-BCR and IgD-BCR, the interaction between BCR and co-receptors, as well as the recruitment of Syk kinase to the BCR signaling subunit Igα. For that, we have chosen the following antibodies (*see* **Note 1**).

   (a) Anti-IgM (clone R33.24.12, in house hybridoma culture; clone 1B4B1, SouthernBiotech);

   (b) Anti-IgD (clone 11-26c.2a, SouthernBiotech);

   (c) Anti-CD19 (clone 6D5; AbD Serotec);

   (d) Anti-CD20 (clone AISB12; eBioscience);

   (e) Anti-Igα (clone JCB117 + HM47/A9 - BSA and Azide free, abcam);

   (f) Anti-Syk (clone Syk-01; BioLegend).

2. Pierce Fab micro preparation kit (Thermo Fisher).

3. Duolink in situ Probemaker plus (Sigma).

4. Duolink in situ Probemaker minus (Sigma).

5. Duolink in situ Detection reagent orange (Sigma, DUO97007) (*see* **Note 2**).

6. Polytetrafluoroethylene (PTFE) 8-well slides (Thermo Scientific Menzel).

7. Sonicator.

8. Milli-Q water.

9. Absolute Ethanol.

10. 1XPBS buffer, room temperature.

11. Kimtech Tissue (Kimberly-Clark professional).

12. 4% Paraformaldehyde (PFA): for 100 mL, Fill a 2000 mL beaker with about 400 mL H_2O and put a thermometer into the beaker. Heat the water on a hot stir plate to 64 °C. Add 98 mL 1XPBS to a 100 mL bottle, add 100 μL of 1 N NaOH and 4 g PFA into the PBS. Put the PBS bottle into the beaker with the warm water and stir the mixture until it is clear. Remove the 4% PFA from the beaker on the hot plate and cool it down to room temperature. Immediately make 1 mL aliquots and freeze at −20 °C. White precipitates are indication for PFA-depolymerization to FA.

13. Permeabilization solution: dissolve 0.5 g Saponin in 100 mL of 1XPBS, filter it through 0.4 μm filter (*see* **Note 3**).

14. Blocking solution: Dissolve 25 mg BSA and 250 μg Sonicated salmon sperm DNA in 100 mL of 1XPBS, filter it through 0.4 μm filter (*see* **Note 3**).

15. Ligation mixture: for 100 μL, mix

5X Duolink ligase buffer	20 μL
Duolink ligase	2.5 μL
1 mM ATP	1 μL
Milli-Q water	76.5 μL

16. Amplification mixture: for 100 μL, mix

Duolink amplification stock 5x	20 μL
Polymerase	1.25 μL
Milli-Q water	78.75 μL

17. DAPI-Fluoromount-G (SouthernBiotech)

3 Methods

3.1 Preparation of the Fab Fragments

Fabs were prepared from the listed antibodies using the Pierce Fab micro preparation kit (Thermo Fisher Scientific), following the manufacturer's instruction.

1. Pack 125 μL of Immobilized Papain settled resins into the 0.8 mL spin column.

2. Wash once with 0.5 mL digestion buffer.

3. Load 125 µg of the antibody to the provided Zeba Spin columns, exchange the buffer to 125 µL Fab-digestion buffer.

4. Load the antibody (in Fab-digestion buffer) to the packed Papain column and digest overnight at 37 °C.

5. The generated Fab fragments were then purified overnight at 4 °C through the provided NAb Protein A Plus Spin columns.

6. The purified Fab fragments were subject to verification by SDS-PAGE and coomassie staining before they can be coupled to the duolink oligos (*see* **Note 4**).

3.2 Couple Fab Fragments with Duolink Oligos

To couple the purified Fabs with duolink oligos, we use the duolink in situ Probemaker plus and minus kit (Sigma-Aldrich) following the manufacturer's instruction.

1. Mix 20 µL of verified Fab-fragments (0.6 mg/mL) (*see* **Note 5**) with 2 µL of conjugation reagent.

2. The mixture was then transferred into glass vials with lyophilized oligonucleotides.

3. Incubate coupling mix at room temperature overnight.

4. Add 2 µL of Stop-solution.

5. Incubate the mixture at room temperature for 30 min to terminate the coupling reaction.

6. Add 24 µL of storage solution to the mixture for stable storage at 4 °C (*see* **Note 6**).

3.3 Clean the PTFE Slides

To reduce background signals (*see* **Note 7**), the PTFE slides have to be cleaned before usage.
To clean the slides

1. Place the slides in a big beaker, cover the slides with absolute ethanol, sonicate for 20 min.

2. Rinse three times with Milli-Q water.

3. Air-dry in a clean box and store at 4 °C.

3.4 Stimulate Cells on Slides

Cultured B cells or splenic B cells isolated from mice were washed with 1XPBS and resuspended into $5-10 \times 10^6$ cells/mL (*see* **Note 8**).

1. Based on the amount of samples (*see* **Note 9**) prepare enough amount of clean PTFE slides (*see* **Note 10**), When using more than one slide, label the slides with a marker pen (*see* **Note 11**).

2. Add 30 µL of the cells to each well of the PTFE slide. Incubate and keep slides in a humid chamber for 30 min at 37 °C (*see* **Note 12**).

3. Verify the attachment of the cells by checking the slides under a microscope (*see* **Note 13**).

4. Suck off PBS with clean Kimtech tissue (*see* **Note 14**).

5. Add 30 µL of stimuli diluted in 1XPBS to the cells to activate the cells for a defined time.

6. Suck off the stimuli with clean Kimtech tissue (*see* **Note 15**).

7. Add 30 µL of 4% PFA to each well, leave the slides at a dark place at room temperature for 20 min to fix the cells (*see* **Note 16**).

8. Wash three times with PBS. To wash the slides, rinse the slides by slowly pipetting 500 µL 1XPBS on the slides. Then place the slides in the humid chamber for 10 min on a shaker platform under slight movement (<20 rpm). Repeat these steps two more times.

9. Suck off PBS with clean Kimtech tissue.

3.5 Permeabilization of the Cells (See Note 17)

1. Add 30 µL of permeabilization solution to each well. Keep the slides at room temperature for 30 min.

2. Wash three times with 1XPBS.

3. Suck off PBS with clean Kimtech tissue.

3.6 Blocking

1. Add 30 µL of blocking solution to each well (*see* **Notes 18 and 19**). Keep the slides at 37 °C for 30 min in the humid chamber.

2. Wash three times with 1XPBS.

3. Suck off PBS with clean Kimtech tissue.

3.7 Amplify Signal and Detection

1. Dilute the plus and minus oligo coupled Fabs with 1XPBS (*see* **Note 20**).

2. Mix equal volume of diluted plus and minus oligo coupled Fabs.

3. Add 30 µL of the mixed oligo coupled Fabs to each well, keep at 4 °C in a dark place overnight (*see* **Note 21**).

4. Remove Fab solutions from each well completely by pipetting, immediately add 30 µL of 1XPBS after the removal of the antibody solution (*see* **Note 22**).

5. Wash three times with 1XPBS.

6. Suck off PBS with clean Kimtech tissue.

7. Add 30 µL of the ligation mixture to each well.

8. Keep the slides at 37 °C for 30 min in the humid chamber.

9. Suck off the ligation mixture with clean Kimtech tissue.

10. Wash three times with 1XPBS.

11. Add 30 µL of the amplification mixture to each well.

12. Keep the slides at 37 °C for less than 100 min (*see* **Note 23**).

13. Suck off the amplification mixture with tissue.

14. Wash two times with duolink buffer B.

15. Wash one time with PBS.

16. Optional step, staining cells with antibodies against other molecules (see **Note 24**).

17. Rinse the slides with absolute ethanol.

18. Keep the slides light protected in a dry chamber and let them dry completely.

19. Add 20 μL DAPI-Fluoromount-G to each well (*see* **Note 25**), cover the slides with coverslip. Keep the slides at 4 °C for at least 15 min before imaging.

3.8 Image Acquisition and Image Analysis

Protein–protein interactions measured by PLA can be detected as discrete fluorescent spots using either a normal fluorescence microscope or a confocal microscope with the appropriate filters for the detection fluorophore used. We suggest using the confocal microscope due to the generally small size of B cells and specially its cytoplasm. It is possible to collect all the PLA signals in one cell by taking stacks of confocal image and reconstruct the 3D image of the cell. However, in most cases, it is enough to collect images containing hundreds of cells from just one focal plane and compare them with images from different samples as long as all images to be compared are acquired in a similar position (*see* **Note 26**).

To compare the signals from different samples, it is also necessary to keep image settings such as laser intensity, exposure time, pixel size, etc. constant during an experiment (*see* **Note 27**).

The collected images can be analyzed by using the Sigma Duolink ImageTool (DUO90806) or the free ImageJ (https://imagej.net/), BlobFinder (http://www.cb.uu.se/~amin/BlobFinder/) or CellProfiler (http://cellprofiler.org) software (*see* **Note 28**).

4 Notes

1. The success of the PLA reaction highly depends on the quality and specificity of the used antibodies. Pure antibodies should come in PBS without any supplement like BSA, carrier proteins, stabilizer, or gelatine. Small amount of azide is tolerable; glycerol and saccharose can be removed by buffer exchange with Zeba-spin columns. Antibodies must be suitable for fixed targets. Whenever it is possible, we would suggest using monoclonal antibodies instead of polyclonal antibodies for better results.

2. Sigma offers Duolink in situ Detection reagent with five options, FarRed, Green, Orange, Red, and BrightField. We often use the Orange for its compatibility with other fluorescent markers we use, such as GFP and its best signal/noise ratio.

3. Small particles or unsolved proteins of BSA and/or FCS interfere with fluorescence microscopy. Using filtered solution would reduce noise and background.

4. Depending on the isotype and subclass of the antibody, it might be better to use Ficin to digest the antibody for Fab preparation. However, it is notable that Ficin digestion of antibody could also produce $F(ab')_2$ if one uses lower concentration of the Cysteine in the reaction (25 mM, pH 5.6 for Fab and 7 mM, pH 5.9 for $F(ab')_2$). To purify the produced Fab, Protein G columns instead of Protein A can be used to remove Fc-parts of antibodies other than mouse IgG2 isotypes such as mouse IgG1 or sheep antibodies.

5. The manufacturer recommends an antibody concentration of 1 mg/mL for coupling with duolink oligos to achieve a roughly 1:1 labeling. For Fab-fragments, a concentration of 0.6 mg/mL up to 0.8 mg/mL is sufficient. Fab fragments of a concentration below 0.5 mg/mL should be concentrated for example using Amicon Ultra—0.5 mL 30 K centrifugal filter units (Merck, UFC503096).

6. The stability of the oligo-coupled Fabs varies depending on many factors. In general, they have only several months of lifetime. We suggest not using any coupled probes that are older than 6 months.

7. PTFE Slides cannot be used directly from the box, even if they are described as "cleaned" or "prewashed" Dust or dirt on the slides will cause nonspecific fluorescence signals under the microscope and finally interfere with real PLA signals.

8. The density of the cells we mentioned here is for the relative small primary B cells and B cell lines with similar size. For larger cells such as certain B-cell lymphoma cell lines, a lower density could be more suitable. The basic requirement is that the cells should not touch each other to avoid the interference of a trans-binding. Moreover, when the cell density is too high, stimuli, detergent, or PLA-probes could remain in the cell–cell interspace and interfere with the results. On the other side, the density of the cells should be high enough for reducing the total amount of microscope images required to achieve significant cell numbers for statistical analysis.

9. We suggest preparing at least double wells for each condition. It is also necessary to have at least two PLA control wells on each slide. We propose to include a positive control with a known interaction to monitor the ligase and polymerase activity. Negative controls can be KO-cells, which do not express the target protein, or the usage of only one of the PLA-probe or the usage of only the detection reagents or the combinations of these conditions.

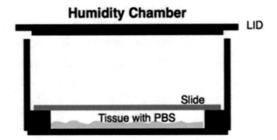

Fig 1 Schematic drawing for the humidity chamber

10. PTFE slides can be autoclaved and used to culture adherent cells. However autoclaving will affect the water repellent surface of the slide. Under this condition, each well can be encircled with a Pap Pen (Thermo Fisher, Sigma-Aldrich) to keep the solutions pooled in a single well. The Pap Pen solution is toxic for all cells; ensure that the circles are completely dried before usage.

11. Each well of the PTFE slide is already labeled by number. One can label the slide at its label area at the bottom of the slide. We strongly suggest preparing an excel table with each condition, interaction, treatment, and PLA probe dilution for each well on each slide to keep the experiment recorded.

12. It is of great importance to never let the slides dry out. A clean tissue saturated with 1XPBS placed on the bottom of a light-tight box can prevent cells from drying out and protect the fluorescence signal from photobleaching (Fig. 1).

13. To avoid the possible activation of the cells directly through the attachment to the slides, we do not coat the slides. By this means, pipetting should be careful and it is necessary to confirm the good attachment of cells to the slides before the experiment starts. Treatment of the slides with poly-lysine or other coatings can facilitate the attachment of the cells. However, it may also stimulate the cells.

14. It is not suggested to remove all the liquid from the well. The residual volume should always be equal, e.g., around 10 μL for each well. Dealing with more than two slides at a time could dry out the sample and will cause high background.

15. For very short stimulation times like 2–10 s, removal of stimuli with Kimtech tissue could be problematic since it could take too long. For that, we suggest immediately fixing the cells by adding 30 μL 8% PFA to the cells directly after the activation.

16. PFA is not the best choice for fixing lipids. To fix the lipids and prevent the potential lateral movement of membrane receptors, one could include 0.02% glutaraldehyde (GA) in the 4% PFA fixing buffer. However, fixing cells with GA could cause

high background fluorescence. For that, one could wash the cells 3 times with 0.5 mg/mL of $NaBH_4$ to reduce aldehyde groups for limiting the background fluorescence.

17. Permeabilization of the cells is only needed for intracellular PLA. Saponin is a mild detergent but can lead to a loss of some parts of the plasma membrane. Thus, permeabilization of the cells with Saponin could change the relative location of membrane protein against intracellular components. For this reason, we do not recommend using PLA to detect interactions across the membrane.

18. Permeabilization of cells can be combined with blocking by including 0.5% of Saponin in the blocking solution.

19. Duolink blocking buffer included in the Sigma kit contains detergent. When the targets are intracellular proteins, one can use it instead of the blocking solution.

20. Dilution of the Fabs is individual and has to be determined by titration experiments. The optimum dilution of the Fabs should allow the detection of 10–20 PLA dots/cell in the positive control and about one PLA dot in every seven cells in the negative control. Fabs can also be diluted with probe-maker assay reagent for intracellular PLA.

21. Fab preparations may contain small amount of the undigested antibodies. Depending on the purity of the Fab preparations, it may be necessary to block the cell surface Fc receptors with Fc blocker to reduce background unspecific signals.

22. Removal of Fab carefully with pipette avoids cross contamination of Fabs between wells. Immediately add 1XPBS to avoid dry-out.

23. Amplification for longer than 100 min would lead to confluent signals, which cannot be quantified anymore.

24. Counterstaining of cellular organelles, membranes, or proteins is possible after the last washing steps. Add 30 μL of the specific antibody with optimized concentration. Incubate overnight at 4 °C. Wash with 1XPBS 3 times and continue with next step (Ethanol rinsing).

25. To quantify PLA signals per cell, it is necessary to stain the cell nucleus for determining the position of cells during data processing. We include DAPI staining in the mounting step. It is possible to stain the cells separately with other nucleus staining dyes. It is necessary to avoid any bubbles when pipetting the Fluoromount G to the cells. Fluoromount G is very viscous. Whenever bubbles are formed, discard the portion that contains the bubbles.

26. It is important to take the images randomly. For that, we only use the signal form nucleus staining to find and focus the cells.

In addition, images should come from all the wells with the same condition to avoid errors caused by the location of the well.

27. To set up the constant parameters for image acquisition, it is necessary to go through all the samples to ensure that the setting will allow low background and still keep the strong signal in the detecting range. It is also important to check each well for abnormal signals that often appear at the edge of the well due to inappropriate handling of the slides such as dry-out. One should not take images in these areas.

28. BlobFinder software is sufficient to analyze PLA experiment results if only a single PLA fluorescence detection is used in addition to the nuclear stain. For PLA experiments with additional fluorescence staining, ImageJ or CellProfiler are the programs of choice.

Acknowledgments

This work was supported by the Excellence Initiative of the German Federal and State Governments (EXC 294), by ERC-grant 322972, and by the Deutsche Forschungsgemeinschaft through TRR130.

References

1. Metzger H (1992) Transmembrane signaling: the joy of aggregation. J Immunol 149:1477–1487

2. Lee S, O'Dowd B, George S (2003) Homo- and hetero-oligomerization of G protein-coupled receptors. Life Sci 74:173–180

3. Lillemeier BF, Pfeiffer JR, Surviladze Z, Wilson BS, Davis MM (2006) Plasma membrane-associated proteins are clustered into islands attached to the cytoskeleton. Proc Natl Acad Sci U S A 103:18992–18997

4. Tao R-H, Maruyama IN (2008) All EGF (ErbB) receptors have preformed homo- and heterodimeric structures in living cells. J Cell Sci 121:3207–3217

5. Yang J, Reth M (2010) Oligomeric organization of the B-cell antigen receptor on resting cells. Nature 467:465–469

6. Thompson MA, Lew MD, Moerner WE (2012) Extending microscopic resolution with single-molecule imaging and active control. Annu Rev Biophys 41:321–342

7. Shivanandan A, Deschout H, Scarselli M, Radenovic A (2014) Challenges in quantitative single molecule localization microscopy. FEBS Lett 588:3595–3602

8. Mattila PK, Feest C, Depoil D, Treanor B, Montaner B, Otipoby KL, Carter R, Justement LB, Bruckbauer A, Batista FD (2013) The actin and tetraspanin networks organize receptor nanoclusters to regulate B cell receptor-mediated signaling. Immunity 38:461–474

9. Avalos AM, Bilate AM, Witte MD, Tai AK, He J, Frushicheva MP, Thill PD, Meyer-Wentrup F, Theile CS, Chakraborty AK, Zhuang X, Ploegh HL (2014) Monovalent engagement of the BCR activates ovalbumin-specific transnuclear B cells. J Exp Med 211:365–379

10. Maity PC, Blount A, Jumaa H, Ronneberger O, Lillemeier BF, Reth M (2015) B cell antigen receptors of the IgM and IgD classes are clustered in different protein islands that are altered during B cell activation. Sci Signal 8:ra93–ra93

11. Lee J, Sengupta P, Brzostowski J, Lippincott-Schwartz J, Pierce SK (2017) The nanoscale spatial organization of B-cell receptors on

immunoglobulin M- and G-expressing human B-cells. Mol Biol Cell 28:511–523

12. Söderberg O, Leuchowius K-J, Gullberg M, Jarvius M, Weibrecht I, Larsson L-G, Landegren U (2008) Characterizing proteins and their interactions in cells and tissues using the in situ proximity ligation assay. Methods 45: 227–232

13. Kläsener K, Maity PC, Hobeika E, Yang J, Reth M (2014) B cell activation involves nanoscale receptor reorganizations and inside-out signaling by Syk. elife 3:e02069

14. Volkmann C, Brings N, Becker M, Hobeika E, Yang J, Reth M (2016) Molecular requirements of the B-cell antigen receptor for sensing monovalent antigens. EMBO J 35:2371–2381

Chapter 13

Single-Particle Tracking of Cell Surface Proteins

Laabiah Wasim and Bebhinn Treanor

Abstract

Single-particle tracking has been used extensively to advance our understanding of the plasma membrane and the mechanisms controlling the movement of cell surface proteins. These studies provide fundamental insights into the regulation of membrane receptor activation and the assembly of signaling clusters. Here, we describe a method to label and track B cell receptor (BCR) and other cell surface proteins and how this method can be adapted to simultaneously track two molecular species or examine the movement of membrane proteins in relation to membrane microdomains. We recently used this method to study the role of the actin cytoskeleton in the regulation of B cell receptor dynamics at the cell surface.

Key words Single-particle tracking, SPT, Membrane proteins, Protein dynamics

1 Introduction

Single-particle tracking (SPT) is a technique used to track the motion of individual proteins or lipids on the surface of cells with a high degree of spatial and temporal resolution [1]. SPT studies continue to advance our understanding of the compartmentalization of the plasma membrane and the structural organization of the underlying actin cytoskeleton. Studying surface proteins by SPT has revealed heterogeneity in protein movement within the membrane, with some exhibiting confined motion whereas others display free diffusion or directed motion [2–6]. These studies provide valuable insights into the mechanisms controlling the movement of surface proteins and how this influences their function, as well as how processes such as receptor crosstalk, endocytosis, and formation of multi-protein complexes occur on a biological timescale. We use SPT to investigate the biophysical mechanisms controlling the dynamics of B cell receptor (BCR) and key activating and inhibitory co-receptors and how this impacts the functional response of B cells. Here, we describe our method to prepare Fab fragments to label BCRs and other cell surface proteins in primary naive splenic B cells and acquire live cell time-lapse images using total internal

Chaohong Liu (ed.), *B Cell Receptor Signaling: Methods and Protocols*, Methods in Molecular Biology, vol. 1707,
https://doi.org/10.1007/978-1-4939-7474-0_13, © Springer Science+Business Media, LLC 2018

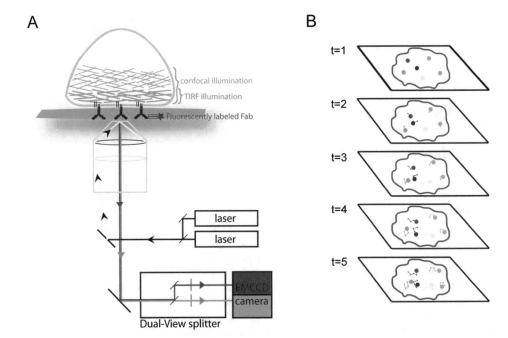

Fig. 1 Schematic diagram of single-particle tracking data acquisition. (**a**) Diagram of microscope instrument for single-particle tracking based on a total internal reflection fluorescence microscope capable of simultaneous acquisition of two fluorophores using a dual-view image splitter and a single EMCCD camera. This can be used to simultaneously visualize and track two different molecular species, or can be used to visualize and track cell surface proteins in relation of the cytoskeleton, lipid microdomains, or macromolecular clusters. (**b**) Schematic diagram of single-particle visualization and tracking over time. *Large dots* indicate position of particle at that timepoint and *small dot* and *linking line* follow particle trajectory at each timepoint. *Red line* outlines area of cell contact

reflection fluorescence microscopy (TIRFM) to track single molecules and derive the diffusion coefficient (Fig. 1). This method can easily be implemented to label and track any cell surface protein in primary cells or cell lines. Moreover, this method can be adapted to simultaneously track (Fig. 1a) two different molecular species, or examine the movement of a cell surface protein in relation to the cytoskeleton, lipid microdomains, or macromolecular complexes.

2   Materials

2.1   Preparation of Fab Fragments

1. IgG antibody.

2. Sample buffer (20 mM sodium phosphate, 10 mM EDTA; pH 7.0).

3. Cysteine-HCl (Sigma).

4. Digestion Buffer (sample buffer +20 mM cysteine-HCl; pH 7.0 *prepare fresh).

5. Papain (Sigma).

6. 1 mM Iodoacetamide (freshly prepared) (Sigma).

7. SDS-PAGE gel.

8. Protein A.

9. Disposable Plastic Column (Pierce).

10. PBS.

11. 50 mM glycine; pH 3.

12. 1 M Tris, pH 8.

13. 2 M Urea.

2.2 Labeling of Fab Fragments for SPT

1. Antibody Fab fragment (at 1–2 mg/ml).

2. 1 M bicarbonate buffer, pH 8.3 (prepare fresh).

3. NHS-ester fluorescent label (recommend Atto-tec 488 or 633; AlexaFluor 488, 555, or 633.

4. PBS.

5. Dialysis cassette (Pierce Slide-a-lyzer).

6. Protein concentration spin column (Amicon, EMD Millipore).

2.3 Preparation of Coated Coverslips for B Cell Adhesion or Activation

1. FCS2 closed chamber system with the temperature controller (Bioptechs, Inc.).

2. 40 mm glass coverslips (Bioptechs, Inc.).

3. Polypropylene scissor-type forceps (Thermo Scientific, Nalgene).

4. Chromic-Sulfuric acid solution (VWR).

5. Acetone.

6. PBS.

7. Anti-MHC Class II antibody (clone M5/114; ATCC TIB120).

8. Fibronectin (Invitrogen).

9. Anti-IgM (Jackson Immunoresearch) or anti-kappa (ATCC HB58).

10. Chamber buffer: PBS, 0.5% fetal bovine serum, 2 mM Mg^{2+}, 0.5 mM Ca^{2+}, 1 μg/ml D-glucose, pH 7.4.

2.4 B Cell Isolation and Labeling

1. Cell strainers, 70 μm (Falcon).

2. Erythrocyte lysis buffer: 0.75% Tris–HCl, 0.2% NH_4Cl, pH 7.2.

3. EasySep™ Mouse B Cell Isolation Kit (Stem Cell).

4. Labeled fab fragment.

5. Chamber buffer: see previous.

2.5   Total Internal Reflection Fluorescence Microscopy for SPT

Single-molecule fluorescence microscopy is performed on a TIRF microscope based on an inverted microscope with a 100x (or higher) NA 1.45 TIRF objective and sensitive EMCCD camera (e.g., Cascade II or Evolve Delta by Photometrics). Simultaneous two-channel acquisition can be achieved by mounting an image splitter (Optosplit II, Cairn Research) in front of the camera port (Fig. 1a) or by mounting two EMCCD cameras and splitting the emission with an appropriate dichroic in the emission light path.

3   Methods

3.1   Preparation of Fab Fragments (See Note 1)

1. If the IgG is purified and lyophilized, proceed to **step 2**. If the IgG is in solution, dialyze against Sample Buffer. Concentrate IgG to approximately 4 mg/ml. Lower concentrations will require a higher enzyme to substrate ratio or a longer digestion time.

2. Just before use, prepare Digestion Buffer by adding cysteine HCl to a final concentration of 20 mM to the Sample Buffer and adjusting the pH to 7.0.

3. Weigh out papain, dissolve 5 mg/ml for soluble papain in digestion buffer, and activate at 37 °C for 10 min.

4. Add papain to antibody, 5 μg soluble papain per 1 mg antibody and react for 4–16 h at 37 °C.

5. Stop the reaction by adding 1 mM (freshly prepared) iodoacetamide and incubate at 4 °C for 30 min.

6. Assay digestion via SDS-PAGE under both reducing and non-reducing conditions followed by Coomassie blue staining (*see* **Note 2**).

7. Purify Fab fragment using Protein A (*see* **Note 3**).

   (a) Prepare 1 ml column of Protein A.

   (b) Wash with PBS.

   (c) Add digested antibody to the column and collect eluate and first 3 ml of PBS wash.

   (d) Add this eluate and wash to the column and collect eluate and first 12 ml of PBS wash in 1 ml fractions (Fab fragment should be in these).

   (e) Elute column with 5 ml 50 mM glycine pH 3 (should be Fc and uncut mAb).

   (f) Neutralize with 1 M Tris, pH 8.

   (g) Wash column with 3 ml 2 M Urea, followed by 1 M LiCl, followed by 50 mM glycine pH 3, followed by thorough washing with PBS.

8. Analyze each fraction on a spectrophotometer at OD280 to estimate protein content, then run SDS-PAGE to identify fractions containing Fab fragment.

9. Concentrate Fab fractions to about 2 mg/ml using a spin column according to the manufacturer's instructions.

3.2 Labeling Fab Fragments for SPT

1. Take a 200 μl aliquot of the Fab fragment (at ~2 mg/ml).

2. Add 20 μl 1 M bicarbonate buffer.

3. Add 2 μl dye (4 mg/ml stock).

4. Vortex briefly, then incubate on a shaker for 1 h at RT.

5. Dialyze against PBS using a slide-a-lyzer cassette (three changes).

6. Collect labeled Fab fragment and analyze on UV-Vis spectrophotometer to assess Fab concentration and ratio of fluorophores per Fab (*see* **Note 4**).

7. Store labeled Fab fragment in a foil-wrapped tube at 4 °C.

3.3 Preparation of Coated Coverslips in FCS2 Chambers (See Note 5)

Glass coverslips are cleaned and coated with either anti-MHC class II antibody or fibronectin to allow for B cell spreading independent of BCR activation or coated with anti-BCR antibodies (anti-IgM or anti-kappa).

1. Submerge glass coverslips (40 mm), held using polypropylene forceps, in chromic-sulfuric acid solution for 20 min, then rinse with water followed by acetone. Allow coverslips to air-dry until acetone has evaporated.

2. Incubate coverslips with 1 ml of 1 μg/ml anti-MHC class II for 4 h at RT (*see* **Note 6**).

3. Wash coverslips with PBS, assemble into FCS2 chambers as per the manufacturer's instructions, and flush with 3 ml chamber buffer for subsequent live imaging.

3.4 Naïve Splenic B Cell Isolation

B cells are purified from murine spleens using negative selection.

1. Homogenize isolated spleen using a cell strainer and syringe plunger (3 ml). The cell strainer is washed with 3 × 3 ml PBS and the single-cell suspension is collected in a 15 ml tube.

2. Resuspend splenocytes in 1 ml of erythrocyte lysis buffer for 5 min at RT and then wash once with 10 ml PBS.

3. EasySep™ Mouse B Cell Isolation Kit is used to purify naïve splenic B cells by negative selection, as per the manufacturer's instructions.

3.5 Labeling B Cell Plasma Membrane Proteins

B cells, purified from murine spleen or cultured cell lines, are labeled at a low concentration of Fab fragment specific for protein of interest to allow visualization and tracking of single particles.

1. Label 5×10^6 primary splenic B cells (or cell line) with fluorescently labeled Fab fragment at an appropriate concentration to allow visualization and detection of single molecules (*see* **Note** 7). For IgM-BCR, label cells with 2 ng/ml anti-IgM Fab fragment in 500 µl chamber buffer at 4 °C for 15 min. This results in labeling of approximately 1 out of every 500 IgM molecules.

2. Wash the cells once with 10 ml PBS and resuspend in 500 µl chamber buffer. Store the samples on ice until acquisition.

3.6 TIRF Microscopy for Detection of Single Particles

1. Insert FCS2 chamber into stage adaptor of TIRF microscope and pre-warm to 37 °C using the temperature controller (Bioptechs, Inc). Incubate the B cell sample at 37 °C in the dark for 5 min prior to image acquisition.

2. Inject B cells into the FCS2 chamber and immediately begin timing. Monitor the settling of cells by brightfield to determine when cells have settled.

3. Acquire images at 20 frames/s for 10 s (50 ms exposure time) at a TIRF penetration depth of ~100 nm above the coverslip. Time-lapse images should be acquired within 10 min after B cells settle on coverslip as the movement of cell surface proteins decreases over time. Analysis of the photobleaching characteristics can be used to confirm that visualized fluorescent puncta are in fact single particles. Single particles labeled with one (or more) fluorophore(s) should display characteristic single-step (or multi-step) photobleaching, as previously described [7].

4. If performing simultaneous two-channel acquisition, image registration can be achieved by measuring the position of fluorescent microspheres (TetraSpek 0.1 µm, Invitrogen).

3.7 Single-Particle Tracking and Analysis

1. Single-particle tracking is performed using the algorithm by Crocker and Grier [8] and implemented in Matlab (MathWorks) by Daniel Blair and Eric Dufresne (http://physics.geo rgetown.edu/matlab/). See Fig. 2 for example. A minimum

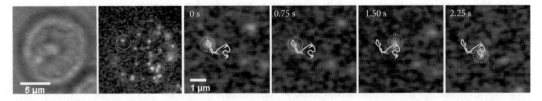

Fig. 2 Single-particle tracking of IgM on the surface of B cells using total internal reflection fluorescence microscopy. Bright-field image (*frame 1*) and TIRF image (*frame 2*) showing single IgM molecules. *Dotted circle* indicates single particle that is magnified in time-lapse. (*Frames 3–6*) Time-lapse of single particle with overlaid track quantified in MATLAB showing diffusion up to 2.25 s

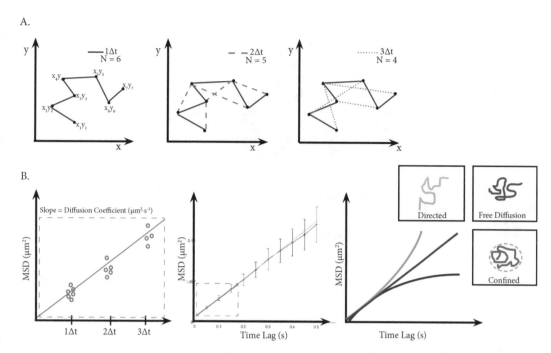

Fig. 3 Calculating diffusion coefficient from two-dimensional single-particle tracking. (**a**) Single-particle undergoing diffusion. (*Left*) Points show position of particle at timepoints separated by the time-lag Δt (s) indicated by *solid lines*. (*Middle*) *Dashed lines* indicate points separated by two time-lags and (*Right*) *dotted lines* indicate points separated by three time-lags. Note that there are less paired points (N) associated with increasing time intervals. Square displacements (μm^2) of pairs of points are calculated and averaged to get the mean square displacement (MSD) for the given time-lag. (**b**) (*Left*) MSD can be plotted with the respective time intervals; the slope of the resulting plot is the diffusion coefficient ($\mu m^2\ s^{-1}$). (*Middle*) MSD vs time-lag graph of experimental data used to calculate a diffusion coefficient of 0.05 $\mu m^2\ s^{-1}$. Diffusion coefficients are calculated using the first three time intervals (indicated by the *dashed box*) as decreased number of paired points for larger time intervals result in greater error. (*Right*) MSD values are directly proportional to their associated time-lags for particles undergoing random or free diffusion. Particles undergoing directed diffusion or diffusion restricted to a confined area have characteristic MSD vs time-lag curves

> track length of 10 was used to discard short tracks and reduce statistical noise.

2. Mean square displacement (MSD) can be calculated for a time lag nΔt using a running average along the trajectory [9] (Fig. 3).

3. Distribution of diffusion coefficients can be plotted as the relative frequency of diffusion coefficients in the indicated bins or as cumulative probability plots (Fig. 4).

4. For the analysis of single-particle tracks in relation of cellular features such as the actin cytoskeleton [7], crop TIRFM images to region of interest and remove noise with a spatial band-pass filter of 1 pixel size. Generate a binary image by thresholding based on Otsu's method [10]. Regions above the threshold identify areas of high fluorescence in the raw image and the mask created can be used to identify sub-trajectories of at least

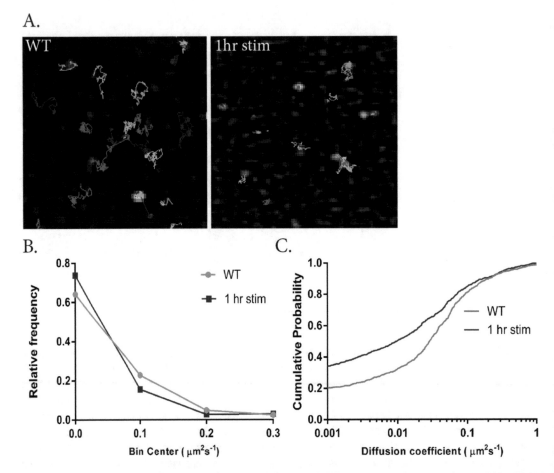

Fig. 4 Distribution of IgM single-particle diffusion coefficient. (**a**) IgM particle tracks in wildtype (WT) and BCR stimulated (1 h stim) B cells quantified in MATLAB. (**b, c**) Quantification of the distribution of diffusion coefficients of IgM molecules in WT and stimulated B cells. (**b**) Relative frequencies of single IgM molecules with diffusion coefficients in the indicated bins. (**c**) Cumulative probability plot of IgM diffusion coefficients. *Left shift of curve* in treated cells indicates higher number of slow diffusing molecules

10 steps inside and outside these regions. MSD analysis can be performed on these regions to determine the diffusion coefficient.

4   Notes

1. If using a commercially prepared unlabeled fab fragment, skip to Subheading 3.2. If using a commercially prepared labeled fab fragment, skip to Subheading 3.3. We have successfully used unlabeled and Cy3-labeled goat anti-mouse IgM and IgG Fab fragments from Jackson Immunoresearch [7].

2. SDS-PAGE and Coomassie blue staining will assess the effectiveness of your antibody digestion. Based on these results you

may need to modify the reaction time, pH, or enzyme concentration. You may want to conduct a pilot time course to assess optimal digestion time.

3. Check the isotype of your antibody for binding affinity to Protein A. Most mouse isotypes bind well to Protein A, with the exception of IgG1. In contrast, rat antibodies do not bind well to Protein A. If necessary, use size exclusion chromatography and DEAE purification instead of Protein A.

4. Be sure to use appropriate calculations for Fab and not whole antibody. 1 mg/ml Fab is OD280 1.4 but MW is ~50 kDa.

5. The method described here uses FCS2 Biotechs chambers and coated-coverslips, but artificial planar lipid bilayers could also be used [11, 12]. For a detailed method on the preparation of artificial planar lipid bilayers, see [13].

6. Cells can alternatively be adhered to coverslips using fibronectin at 1–4 μg/ml for 1 h. Note: concentration of fibronectin will have to be determined for specific cell line.

7. Concentration of Fab fragment used for labeling must be empirically determined for each surface protein of interest to achieve a labeling density in which single particles can be clearly distinguished under TIRF microscopy using SPT imaging conditions. The localization and tracking algorithms will not accurately track if the labeling density is too high, and conversely, a labeling density that is very low will result in too few tracks for each individual cell. One should achieve a labeling density that permits a minimum of 50 tracks per cell.

References

1. Saxton MJ, Jacobson K (1997) Single-particle tracking: applications to membrane dynamics. Annu Rev Biophys Biomol Struct 26:373–399

2. Andrews NL, Lidke KA, Pfeiffer JR, Burns AR, Wilson BS, Oliver JM, Lidke DS (2008) Actin restricts FcepsilonRI diffusion and facilitates antigen-induced receptor immobilization. Nat Cell Biol 10:955–963

3. Espenel C, Margeat E, Dosset P, Arduise C, le Grimellec C, Royer CA, Boucheix C, Rubinstein E, Milhiet PE (2008) Single-molecule analysis of CD9 dynamics and partitioning reveals multiple modes of interaction in the tetraspanin web. J Cell Biol 182:765–776

4. Fujiwara T, Ritchie K, Murakoshi H, Jacobson K, Kusumi A (2002) Phospholipids undergo hop diffusion in compartmentalized cell membrane. J Cell Biol 157:1071–1081

5. Edidin M, Zuniga MC, Sheetz MP (1994) Truncation mutants define and locate cytoplasmic barriers to lateral mobility of membrane glycoproteins. Proc Natl Acad Sci U S A 91:3378–3382

6. Jaqaman K, Kuwata H, Touret N, Collins R, Trimble WS, Danuser G, Grinstein S (2011) Cytoskeletal control of CD36 diffusion promotes its receptor and signaling function. Cell 146:593–606

7. Treanor B, Depoil D, Gonzalez-Granja A, Barral P, Weber M, Dushek O, Bruckbauer A, Batista FD (2010) The membrane skeleton controls diffusion dynamics and signaling through the B cell receptor. Immunity 32:187–199

8. Crocker JC, Grier DG (1996) Methods of digital video microscopy for colloidal studies. J Colloid Interface Sci 179:298–310

9. Ritchie K, Shan XY, Kondo J, Iwasawa K, Fujiwara T, Kusumi A (2005) Detection of non-Brownian diffusion in the cell membrane

in single molecule tracking. Biophys J 88:2266–2277

10. Otsu N (1979) A threshold selection method from gray-level histograms. IEEE Trans Sys Man Cyber 9:62–66

11. Liu W, Meckel T, Tolar P, Sohn HW, Pierce SK (2010) Intrinsic properties of immunoglobulin IgG1 isotype-switched B cell receptors promote microclustering and the initiation of signaling. Immunity 32:778–789

12. Tolar P, Hanna J, Krueger PD, Pierce SK (2009) The constant region of the membrane immunoglobulin mediates B cell-receptor clustering and signaling in response to membrane antigens. Immunity 30:44–55

13. Fleire SJ, Batista FD (2009) Studying cell-to-cell interactions: an easy method of tethering ligands on artificial membranes. Methods Mol Biol 462:145–154

Chapter 14

The Use of Intravital Two-Photon and Thick Section Confocal Imaging to Analyze B Lymphocyte Trafficking in Lymph Nodes and Spleen

Chung Park, Il-Young Hwang, and John H. Kehrl

Abstract

Intravital two-photon laser scanning microscopy (TP-LSM) has allowed the direct observation of immune cells in intact organs of living animals. In the B cell biology field TP-LSM has detailed the movement of B cells in high endothelial venules and during their transmigration into lymph organs; described the movement and positioning of B cells within lymphoid organs; outlined the mechanisms by which antigen is delivered to B cells; observed B cell interacting with T cells, other cell types, and even with pathogens; and delineated the egress of B cells from the lymph node (LN) parenchyma into the efferent lymphatics. As the quality of TP-LSM improves and as new fluorescent probes become available additional insights into B cell behavior and function await new investigations. Yet intravital TP-LSM has some disadvantages including a lower resolution than standard confocal microscopy, a narrow imaging window, and a shallow depth of imaging. We have found that supplementing intravital TP-LSM with conventional confocal microscopy using thick LN sections helps to overcome some of these shortcomings. Here, we describe procedures for visualizing the behavior and trafficking of fluorescently labeled, adoptively transferred antigen-activated B cells within the inguinal LN of live mice using two-photon microscopy. Also, we introduce procedures for fixed thick section imaging using standard confocal microscopy, which allows imaging of fluorescently labeled cells deep in the LN cortex and in the spleen with high resolution.

Key words B cell, CD62L, High endothelial venules (HEVs), Inguinal lymph node, Anti-IgM, Intravital imaging, Two-photon microscopy, Thick section imaging

---

1 Introduction

Blood-borne B cells do not have to cross an endothelial barrier to enter the spleen in order to populate the splenic B cell follicles. B cells enter into the splenic white pulp via terminal arterioles that branch from the central arterioles that frequently end in the marginal sinus, which surrounds the splenic white pulp. CXCL13 is needed for B cells to migrate from the marginal sinus and the

Chaohong Liu (ed.), *B Cell Receptor Signaling: Methods and Protocols*, Methods in Molecular Biology, vol. 1707,
https://doi.org/10.1007/978-1-4939-7474-0_14, © Springer Science+Business Media, LLC 2018

bridging channels into the white pulp B cell follicles [1]. Blood-borne B cells homing into a LN to enter a follicle have a more difficult task. First, they must be captured by high endothelial venules (HEVs), specialized endothelial cells in LNs; next, they must cross the blood vessel endothelium via either para- or trans-cellular migration; and, finally, they must negotiate the surrounding pericytes layer. Only then can they enter into the LN parenchyma and eventually the follicle. To tether and roll along the HEV luminal surface B cells use CD62L, which binds carbo-hydrate ligands recognized by the monoclonal antibody MECA-79 [2, 3]. Specifically, CD62L interacts with sulfated sialomucin ligands such as CD34 and GlyCAM-1 on LN HEVs and ligands located on mucosal addressin cell adhesion molecule-1 (MadCAM 1) in Peyer's patch HEVs [4, 5]. Underscoring the importance of CD62L for B cell homing, B cells poorly populate LNs in CD62L deficient mice [6], and treatment of mice with a CD62L blocking antibody inhibits the entrance of B cells [7]. Furthermore, activation of metalloproteinase-dependent shedding of CD62L can lower its expression, thereby impairing B cell LN homing. ADAM17 (TACE) is the major membrane-associated metallopro-tease responsible [8]. After L-selectin mediated capture the rolling B cells engage chemokines also localized on the luminal surface of the HEV. CCL19/CCR7 and CXCL12/CXCR4 interactions direct B cells into LNs, while CXCL13/CXCR5 interactions direct B cells into Peyer's patches. The engagement of chemokine receptors triggers the activation of integrins. Adhesive filopodia expressing high affinity LFA-1 help B cells search for HEV TEM sites, and once located they assist B cell transmigration by engaging subluminal integrin ligands [9]. After crossing the endo-thelial basement membrane B cells must still negotiate the perive-nule spaces created by overlapping pericytes. Finally, across all barriers LN B cells reach the paracortical cords that originate in-between and below LN follicles and extend into the medullary region [10–12]. B cells migrate along CCL19/21 expressing fibroreticular cells toward the LN follicle where CXCR5 expres-sion directs them into the CXCL13 rich LN follicle [13–15]. Newly resident B cells move toward the follicle center, while longer-term residents tend toward the follicle edges closer to nearby LN egress sites [16, 17].

The use of two-photon laser scanning microscopy (TP-LSM) technology has allowed the monitoring in live animals of adoptively transferred, fluorescently labeled lymphocytes predominately in LNs, but also at other in vivo sites. By utilizing animals with endogenous fluorescently tagged cells, injecting fluorescently labeled antigen, and by injecting fluorescently labeled antibodies directed at other cell types into the lymph or blood, the relationship of transferred lymphocytes to their microenvironment can be assessed [18–21]. Initial studies examining the homing of lymphocytes through HEVs used intravital bright field and epifluorescence video microscopy to study the mechanics of lymphocyte TEM while more recent studies have employed TP-LSM [22–24]. Yet the use of intravital imaging to assess these events in the intact LN is limited to superficial regions of the LN. Combining conventional immunohistochemistry using thick LN sections with intravital TP-LSM can provide a more detailed view of B cell trafficking and immune cell interactions. Here, we review our methods for combining TP-LSM imaging [25, 26] with thick section LN and spleen imaging [27]. We outline these methods and demonstrate their use in the study of the trafficking of antigen activate mouse B cells. We compare the behavior of recently activated cells to that of similar cells not stimulated via their antigen receptor. One advantage of the adoptive transfer approach used here is that both the experimental and control cells can be imaged in parallel in a single animal using their different fluorescent labels. Together these studies demonstrated that antigen-activated B cells enter lymph nodes poorly due to a failure of the activated B cells to be captured by HEVs. The likely mechanism is antigen receptor triggered activation of Adam17 followed by CD62L shedding. In contrast, the activated B cells readily enter the splenic white pulp. Many of the recently transferred control and activated cells in the spleen can be found in the bridging channels, a site where the newly arrived B cells can meet antigen-bearing dendritic cells.

2 Materials

2.1 Preparation of Lymphocytes from Mouse Spleen

1. 1× Phosphate-buffered saline (PBS), pH 7.4.

2. RPMI-1640 medium (Gibco).

3. Fetal calf serum (FCS), quantified (Gibco).

4. 40 and 100 μm of Nylon Cell strainer (BD Falcon).

5. 25 × 5/8—gage needles (Kendall).

6. 12 mL syringe (Kendall).

7. ACK lysis buffer (BioWhittaker, Lonza).

8. Bovine serum albumin (BSA) fraction V.

9. Biotinylated mouse-Antibodies (MAb) (BD Parmingen™).

10. Magnetic negative selection systems: Dynabeads and magnetic particle concentrator (Invitrogen).

11. Fluorescence-activated cell sorting (FACS) buffer: 1 × PBS, pH 7.4, 1% BSA.

2.2 Labeling Lymphocytes and In Vivo Transfer of Labeled-Lymphocytes

1. CellTracker™ probes (Molecular Probes, Invitrogen) for long-term tracing of living cells.

2. 5-Chloromethylfluorescein diacetate (CMFDA) and 5-(and-6)-(((4-chloromethyl)benzoyl)amino)tetramethylrhodamine (CMTMR), stock solution 5 mM in DMSO (stored at −20 °C).

3. 5 mM of eFluor450 in DMSO (eBioscience).

4. RPMI-1640 media (Gibco).

5. Fetal calf serum (FCS), qualified (Gibco).

6. 1× Phosphate-buffered saline (PBS), pH 7.4.

2.3 Visualization of HEVs in LN

1. Evans Blue solution (0.5 µg/mL in PBS) (Sigma).

2. Purified Rat anti-mouse PNAd carbohydrate epitope (BD Pharmingen™, clone: MECA-79) for HEVs.

3. Antibody Labeling Kits for fluorescents (Alexa Fluor® 488 or 647) (Cat# A20181, A20185, or A20189, Molecular probe) for fluorescents labeling.

4. 1× Phosphate-buffered saline (PBS), pH 7.4.

2.4 Multi-photon Imaging

1. Two-photon laser-scanning microscopy system. We use a Leica SP8 inverted 5 channel confocal microscope, a Ti:Sapphire laser (Spectra Physics) with a 10 W pump tuned to 810 nm, a 25× water dipping objective with long working distance (for example, Leica 25× 0.95NA), an incubation cube chamber (Life Imaging Services, Basel, Switzerland) (*see* **Note 1** and Fig. 1a).

2. Heating Mantles & Blankets (Thermo Scientific). The heating blanket is set up to heat at the 37 °C.

3. Anesthesia: 50× Avertin (tribromoethanol and 2-Methyl-2-butanol, Sigma-Aldrich), stored at −20 °C. Dilute prior to use as 1×; *see* later.

4. Small animal clipper (Fisher Scientific).

5. Microsurgical instruments: Roboz surgical scissor and forceps.

6. Cover-glass chamber slide (Nalgene, Nunc).

7. Classic Double Edge Blades 5-Pack (Wilkinson Sword).

8. 1× Phosphate-buffered saline (PBS), pH 7.4.

9. Cyanoacrylate glue.

10. Infusion set (Butterfly ST, 25× 3/8 32/2″ Tubing).

2.5 Thick Section of LN and Spleen Preparation and Confocal Microscopy

1. Vibratome (Leica VT-1000 S) (*see* Fig. 1e).

2. Leica SP8 inverted 5 channel confocal microscope.

3. Paraformaldehyde (Electron Microscopy Sciences).

4. Anti-CD21/35 (BioLegend), anti-B220 (RA3-6B2, BD Biosciences), anti-CD11c (HL3, BD Biosciences), anti-CD169

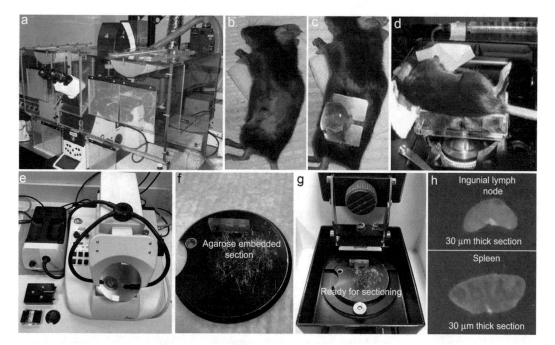

Fig. 1 A two-photon laser-scanning microscopy system and an example of the results of surgery to expose the inguinal LN, and the preparation of thick sections by vibratome. (**a**) Two-photon laser-scanning microscopy system including a Leica SP5 inverted 5 channel confocal microscope. The cube and box, cube chamber, maintains temperature at 37.0 °C ± 0.5 °C. (**b**) Anesthetized mouse is shaved and has small incision on its flank. (**c**) Immobilized inguinal LN with cyanoacrylate glue on blade fitted to a chamber slide. To protect tissue from drying periodically add a few drops of PBS on exposed inguinal LN. (**d**) Complete setup for imaging is shown. Mouse is placed in a pre-warmed chamber slide. Blade can hold the exposed inguinal LN to minimize vibrations caused by breathing. Mouse is kept anesthetizing through intraperitoneal injection of Avertin using an infusion set. (**e**) A vibratome (Leica VT-1000 S). (**f**) An agarose block of spleen and LNs which was embedded with 4% low melting agarose mounted on a cutting stage with a cyanoacrylate containing adhesive (Instant Krazy glue). (**g**) Complete setup to cut an lymphoid organ embedded agarose block. (**h**) An example of a thick section prepared from the agarose block. The lymph node was cut longitudinally while a cross-section was made of the spleen

(3D6.112, BioLegend), Purified Rat anti-mouse PNAd carbohydrate epitope (BD Pharmingen™, clone: MECA-79) for HEVs.

5. 1× Phosphate-buffered saline (PBS), pH 7.4.

6. A cyanoacrylate containing adhesive (Instant Krazy glue).

3 Methods

Here, we describe details of procedures for real-time imaging of adoptively transferred lymphocytes visualized with fluorescent dyes in the inguinal LN of mice. We also describe how to visualize the LN microenvironments. In addition, we introduce procedures for thick section imaging that allow further observation with the imaged organ or the other organs in same mouse, which can be collected after the TP-LSM imaging. This combination method helps the appreciation of serial events in the intact organ of a live animal and can provide further details in the deep LN cortex.

3.1 Preparation of B Cells from Mouse Spleen (See Note 2) [28]

1. Sacrifice mice and harvest the spleens.

2. The spleens are disrupted with forceps or needles in PBS and the cells dissociated by gentle teasing followed by filtering through a 100 and then a 40 μm nylon cell strainer to remove connective tissue cells.

3. Splenocytes are depleted of red blood cells with ACK buffer and subsequently washed with PBS.

4. Centrifuge the cell suspensions at $350 \times g$ for 5 min at 4 °C, remove the supernatants, and resuspend the pellet in FACS buffer.

5. A count of the suspensions is made to determine cell number per milliliter (mL).

6. Pellet the cells and resuspend in FACS buffer. Stain the cells with biotinylated monoclonal antibodies (MAb) to non-B-cell markers such as CD4, CD8, GR-1, Mac-1, Terl19, CD11c, and DX5 in FACS buffer (see **Note 3**).

7. Incubate the cells at 4 °C for 15 min and wash with FACS buffer.

8. Resuspend the pelleted cells in FACS buffer. Wash the Dyna-beads M-280 streptavidin beads with PBS twice and with FACS buffer twice. The bead-to-cell ratio is determined by the manufacturer's protocols [29].

9. Add the washed magnetic beads to the cell suspensions and incubate at 4 °C, slowly rotate for 15 min. The suspensions are then attached to the magnet and allowed to separate.

10. The non-adherent suspension is collected and reapplied to the magnetic source, and again the non-adherent cells are collected: This population is washed, counted, resuspended in RPMI/10% FCS, and incubated at 37 °C for 30 min prior to any stain.

3.2 Labeling Lymphocytes and Their Adoptive Transfer

1. Resuspend $3 \sim 10 \times 10^6$ B cells/mL in RPMI/10% FCS with $1 \sim 5$ μM of CMFDA and CMTMR [30] or 5 μM of eFluor 450, respectively. Incubate the lymphocyte for 15 min at 37 °C and wash five times with PBS (see **Note 4**).

2. Resuspend labeled B cells in PBS and inject intravenously into lateral tail vein of recipient mice.

3. To visualize TEM in HEVs intravital imaging is best performed immediately after cell injection.

3.3 Visualization of Microvessels Including HEVs in the Inguinal LN

1. To visualize microvessels inject 50 μL of Evans Blue solution (0.5 μg/mL in PBS) (Sigma) into the tail or orbital vein.

2. To visualize HEVs using a direct marker inject 20 μL of fluorescently labeled anti-mouse PNAd (MECA-79) (0.25 mg/mL in PBS) (Sigma) into tail or orbital vein.

3.4 Anesthesia

1. Make 50× Avertin (76.9%) stock by dissolving 5 g Tribromoethanol in 6.5 mL of 2-Methyl-2-butanol.

2. The solution must be filtered with a 0.22 μm syringe filter (Millipore) before making aliquots of the solution in 1 mL stock vials. It can be maintained at −20 °C for 1 year.

3. To make 1× Avertin (1.51%), mix 200 μL of 50× Avertin stock with the 10 mL pre-warmed PBS. This solution can be stored at 4 °C for 2 weeks.

4. Inject the mouse, intraperitoneal, with 1× Avertin at a dose of 300 mg/kg. Check the level of anesthesia by monitoring body and respiratory function [19].

3.5 Operation to Expose the Inguinal LN (ILN)

For all of the following steps the mouse should be transferred onto the heated blanket. Inguinal LNs are prepared for intravital microscopy using a modification of a published method [25, 26].

1. After shaving the hair of the left or right flank with a small animal clipper, remove the skin and fatty tissue over the inguinal LN (see Fig. 1b).

2. The exposed inguinal LN of mouse is held with blade fitted to a coverglass chamber slide (see Fig. 1c).

3. Place the mouse in a pre-warmed coverglass chamber slide and add 2 mL of pre-warmed PBS, 37 °C (see **Note 5** and Fig. 1d).

4. Place the chamber slide into the temperature control cube chamber on Leica SP8 microscope systems or equivalent (see **Notes 1** and **5**, and Fig. 1d). Distilled water is used as an immersion fluid.

5. To visualize HEVs, image over the interfollicular channel, between LN follicles (where nearby blood vessels are located) or at the interface between B and T cell zones (underneath the LN follicle).

3.6 Imaging Using Multi-photon Microscopy (See Fig. 1)

1. Tune the Mai Tai Ti:Sapphire laser to 790 ~ 910 nm.

2. For four-dimensional analysis of cell migration, set up the microscope configuration and filter set for the detection of the following dyes (Alexa Fluor® 488, 594, 647, CMFDA, CMTMR, eFluor 450, and second harmonics) (see **Note 6**).

3. Using mercury lamp illumination look through the eyepiece focusing onto the surface of the inguinal LN.

4. Start scanning using the Mai Tai Ti:Sapphire laser (*see* **Note 7**). Adjust the gain and offset of the detectors. Typically, labeled HEVs can be found in the interfollicular channels or at the bottom of the LN follicles. While scanning move the x-, y-, and z-axes to locate an area where a sufficient signal is detected with specific markers for HEVs.

5. Once a suitable location is found, acquire z stacks of optical sections depending on the purpose of the experiment. To examine lymphocyte adhesion and TEM on HEV set a volume that can be scanned every 1 or 2 s; for lymphocytes exiting through the lymphatic endothelium cortex to the lymphatic sinus set a volume that can be scanned every 5 or 10 s.

6. Check the level of anesthesia before starting a new image acquisition.

3.7 Analysis

1. Transform sequence of image stacks into volume-rendered four-dimensional movies using Imaris software (Bitplane) or similar software (*see* Fig. 2a).

2. Use the spot analysis in Imaris for semi-automated tracking of cell motility in three dimensions using the parameters for auto-regressive motion. The tracks of interested must be verified and corrected manually.

3. Calculation of cell motility parameters (track speed, speed variability, straightness, length, and displacement) can be performed using Imaris or other related software programs.

4. The specific pattern analysis of TEM on HEV can be performed using the spot function in Imaris. On the basis of typical changes in cellular morphology that occur during TEM the duration of TEM of individual cells can be measured.

5. Statistical analysis and significance of statistics calculation can be performed with GraphPad Prism (GraphPad Software) or related software programs.

6. QuickTime movies can be generated using a maximum intensity projection of each Z stack. Editing the movies can be performed with Adobe Premier Pro CS3 program (Adobe).

3.8 Thick Section Imaging

1. Fix isolated LNs or spleens in freshly prepared 4% paraformaldehyde (Electron Microscopy Science) overnight at 4 °C on an agitation stage.

2. Embed spleens or LNs with 4% low melting agarose (Invitrogen) in PBS and section with a vibratome (Leica VT-1000 S) at a 30 μm thickness.

3. Block thick sections in PBS containing 10% fetal calf serum, 1 mg/mL anti-Fcγ receptor (BD Biosciences), and 0.1% Triton X-100 (Sigma) for 30 min at room temperature.

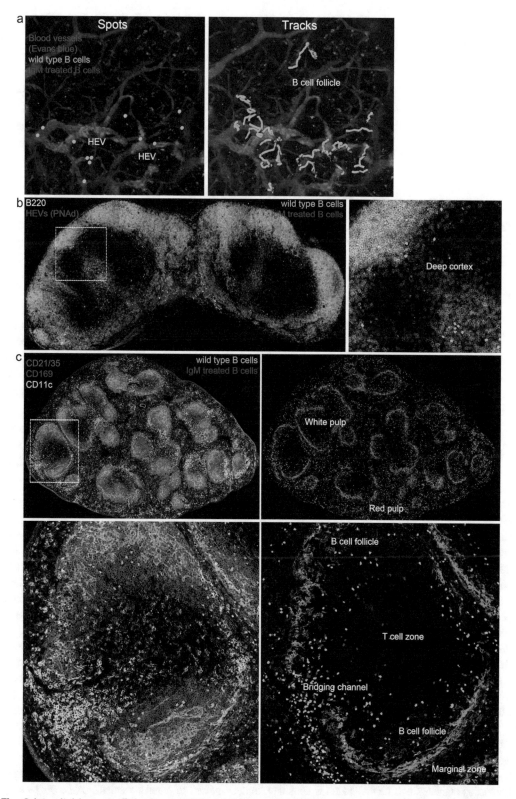

Fig. 2 Intravital images of the inguinal LN and thick section images of the LN and spleen. (**a**) Shown are maximum projections of a z stack (200 μm, 6 μm slices). The blood vessels were visualized by the intravenous

4. Stain sections overnight at 4 °C on an agitation stage with the following antibodies: anti-CD21/35 (BioLegend), anti-B220 (RA3-6B2, BD Biosciences), anti-CD11c (HL3, BD Biosciences), anti-CD169 (3D6.112, BioLegend), anti-mouse PNAd carbohydrate epitope (BD Pharmingen™, clone: MECA-79) (*see* Fig. 2b, c).

5. Analyze stained thick sections using a Leica SP8 confocal microscope (Leica Microsystem, Inc.) and for whole area of section process images with tile function of Leica LAS AF software (Leica Microsystem, Inc.) and optimize image with Imaris software (Bitplane AG).

3.9 FACS Analysis   The conventional FACS protocol can be applied to measure receptor expression level on B cells (*see* Fig. 3a, b).

4 Notes

1. It is recommended the non-descanned detectors be used for collecting even very weak fluorescent signals. The air temperature in the cube chamber (The Cube and Box) (Life Imaging Services) should be monitored and maintained at 37.0 °C ± 0.5 °C. Temperature control is very important for detecting proper cell movement [4].

2. B lymphocytes can be purified from bulk population by removing non-B-lymphocytes. There are multiple methods to achieve this purification; this subheading describes one such procedure

◀ ───

Fig. 2 (continued) injection of Evans *Blue* (*purple*) and the HEVs were outlined by adherent B cells. Fluorescently labeled wild-type B cells (*green*) and anti-IgM (10 μg/mL, 5 h) treated B cells (*red*) were adoptively transferred just prior to imaging. The HEVs and LN follicles are indicated. Spots (*left panel*) and tracks (*right panel*) were generated with functions in the Imaris software. (**b**) Thick section images of inguinal LN previously intravital imaged intravitally with TP-LSM. The LN was fixed and sliced at a 30 μm thickness. Thick sections were immunostained with the indicated markers: B220 (*white*) and PNAd (*purple*). Fluorescently labeled wild-type B cells (*green*) and anti-IgM treated B cells (*red*) are shown in the section. A sagittal section of the inguinal LN was scanned with Leica SP8 confocal microscopy using the tile function to visualize the entire LN section. *Dotted line box* in the left panel was electronically zoomed and shown in the right panel. Very few activated B cells have gained entrance to the LN follicles and deep cortex. (**c**) Thick section image of the spleen. The spleen from the mouse used for TP-LSM was fixed and sections prepared (30 μm thickness slices). The sections were immunostained for the indicated markers: CD21/35 (*blue*), CD169 (*purple*), and CD11c (*yellow*). Fluorescently labeled wild-type B cells (*green*) and anti-IgM treated B cells (*red*) can be seen in the section. Five different signals are shown together in the upper left panel. The *dotted line box* in the upper left panel was electronically zoomed and is shown below in the lower left panel. The CD21/35 and CD11c signals were omitted in the adjacent right panels to better allow visualization of the transferred B cells. The border between *white pulp* and *red pulp* can be delineated by marginal zone CD169 positive macrophages. Numerous activated and non-activated B cells can be seen poised to enter the splenic *white pulp* via the bridging channel (lower right).

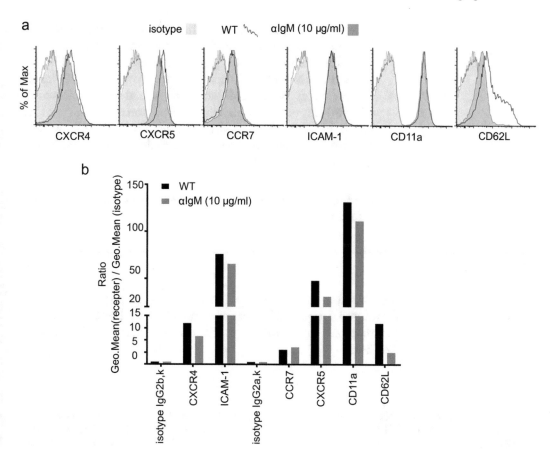

Fig. 3 FACS analysis of surface molecule on B cells treated with anti−IgM or not. Aliquots of B cells to be adoptively transferred into mice were analyzed by flow cytometry for the indicated markers. (**a**) Histogram graphs showing the comparison between anti-IgM (10 μg/mL, 5 h) treated B cells versus B cells cultured in the absence of stimulating antibody. (**b**) *Bar graph* showing the mean fluorescent intensity of various markers divided by the mean fluorescence intensity of the appropriate isotype control antibody. Results are from part **a**. Note the reduction in CD62L expression following a brief exposure to anti-IgM

(refer to the methods published by Dynal, Inc., Miltenyi Biotec-MACS).

3. These biotinylated MAbs will bind CD4 & CD8 T cells, macrophages, erythrocytes, NK cells, granulocytes, and dendritic cells; or will bind B cells, macrophages, erythrocytes, NK cells, granulocytes, and dendritic cells.

4. When staining low cell concentrations of cells FCS can protect cells from the toxic effect of the concentrated dye.

5. Inguinal LN is maintained a physiological temperature because lymphocyte motility and behavior are highly temperature dependent [18].

6. Wavelength separation was through a dichroic mirror at 560 nm and then separated again through a dichroic mirror at 495 nm followed by 525/50 emission filter for CMFDA or Alexa Fluor® 488 (Molecular probes); and the eFluor450 (eBioscience) or second harmonic signal was collected by 460/50 nm emission filter; a dichroic mirror at 650 nm followed by 610/60 nm emission filter for CMTMR or Alexa Fluor® 594; and the Evans blue or Alexa Fluor® 647 signal was collected by 680/50 nm emission filter.

7. Scrupulous care should be taken to minimize the laser power used for excitation as excessive power can cause phototoxicity resulting in cell movement arrest [31].

Acknowledgments

The authors would like to thank Dr. Anthony Fauci for his continued support. This research was supported by the intramural program of the National Institutes of Allergy and Infectious Diseases.

References

1. Mebius RE, Kraal G (2005) Structure and function of the spleen. Nat Rev Immunol 5 (8):606–616

2. Butcher EC, Picker LJ (1996) Lymphocyte homing and homeostasis. Science 272:60–66

3. Rosen SD (2004) Ligands for L-selectin: homing, inflammation, and beyond. Annu Rev Immunol 22:129–156

4. Berg EL, Mullowney AT, Andrew DP, Goldberg JE, Butcher EC (1998) Complexity and differential expression of carbohydrate epitopes associated with L-selectin recognition of high endothelial venules. Am J Pathol Feb 152 (2):469–477

5. Berg EL, McEvoy LM, Berlin C, Bargatze RF, Butcher EC (1993) L-selectin-mediated lymphocyte rolling on MAdCAM-1. Nature 366 (6456):695–698

6. Steeber DA, Green NE, Sato S, Tedder TF (1996) Lyphocyte migration in L-selectin-deficient mice. Altered subset migration and aging of the immune system. J Immunol 157 (3):1096–1106

7. Gallatin WM, Weissman IL, Butcher EC (1983) A cell-surface molecule involved in organ-specific homing of lymphocytes. Nature 304(5921):30–34

8. Peschon JJ, Slack JL, Reddy P, Stocking KL, Sunnarborg SW, Lee DC, Russell WE, Castner BJ, Johnson RS, Fitzner JN, Boyce RW, Nelson N, Kozlosky CJ, Wolfson MF, Rauch CT, Cerretti DP, Paxton RJ, March CJ, Black RA (1998) An essential role for ectodomain shedding in mammalian development. Science 282(5392):1281–1284

9. Shulman Z, Shinder V, Klein E et al (2009) Lymphocyte crawling and transendothelial migration require chemokine triggering of high-affinity LFA-1 integrin. Immunity 30:384–396

10. Gretz JE, Anderson AO, Shaw S (1997) Cords, channels, corridors and conduits: critical architectural elements facilitating cell interactions in the lymph node cortex. Immunol Rev 156:11–24

11. Willard-Mack CL (2006) Normal structure, function, and histology of LNs. Toxicol Pathol 34:409–424

12. Ma B, Jablonska J, Lindenmaier W, Dittmar KE (2007) Immunohistochemical study of the reticular and vascular network of mouse lymph node using vibratome sections. Acta Histochem 109:15–28

13. Bajénoff M, Egen JG, Koo LY et al (2006) Stromal cell networks regulate lymphocyte entry, migration, and territoriality in lymph nodes. Immunity 25:989–1001

14. Förster R, Mattis AE, Kremmer E et al (1996) A putative chemokine receptor, BLR1, directs B cell migration to defined lymphoid organs and specific anatomic compartments of the spleen. Cell 87:1037–1047

15. Reif K, Ekland EH, Ohl L et al (2002) Balanced responsiveness to chemoattractants from adjacent zones determines B-cell position. Nature 416:94–99

16. Park C, Hwang IY, Sinha RK et al (2012) Lymph node B lymphocyte trafficking is constrained by anatomy and highly dependent upon chemoattractant desensitization. Blood 119:978–989

17. Sinha RK, Park C, Hwang IY et al (2009) B lymphocytes exit lymph nodes through cortical lymphatic sinusoids by a mechanism independent of sphingosine-1-phosphate-mediated chemotaxis. Immunity 30:434–446

18. So PT, Dong CY, Masters BR, Berland KM (2000) Two-photon excitation fluorescence microscopy. Annu Rev Biomed Eng 2:399–429

19. Miller MJ, Wei SH, Parker I, Cahalan MD (2002) Two-photon imaging of lymphocyte motility and antigen response in intact lymph node. Science 296:1869–1873

20. Miller MJ, Wei SH, Cahalan MD, Parker I (2003) Autonomous T cell trafficking examined in vivo with intravital two-photon microscopy. Proc Natl Acad Sci U S A 100:2604–2609

21. von Andrian UH, Mempel TR (2003) Homing and cellular traffic in lymph nodes. Nat Rev Immunol 3:867–878

22. von Andrian UH (1996) Intravital microscopy of the peripheral lymph node microcirculation in mice. Microcirculation 3:287–300

23. Park EJ, Peixoto A, Imai Y et al (2010) Distinct roles for LFA-1 affinity regulation during T-cell adhesion, diapedesis, and interstitial migration in lymph nodes. Blood 115:1572–1581

24. Boscacci RT, Pfeiffer F, Gollmer K et al (2010) Comprehensive analysis of lymph node stroma-expressed Ig superfamily members reveals redundant and nonredundant roles for ICAM-1, ICAM-2, and VCAM-1 in lymphocyte homing. Blood 116:915–925

25. Han SB, Moratz C, Huang NN et al (2005) Rgs1 and Gnai2 regulate the entrance of B lymphocytes into lymph nodes and B cell motility within lymph node follicles. Immunity 22:343–354

26. Park C, Hwang IY, Kehrl JH (2009) Intravital two-photon imaging of adoptively transferred B lymphocytes in inguinal lymph nodes. Methods Mol Biol 571:199–207

27. Park C, Arthos J, Cicala C, Kehrl JH (2015) The HIV-1 envelope protein gp120 is captured and displayed for B cell recognition by SIGN-R1(+) lymph node macrophages. eLife 4: e06467. https://doi.org/10.7554/eLife. 06467

28. Moratz C, Kehrl JH (2004) In vitro and in vivo assays of B-lymphocyte migration. Methods Mol Biol 271:161–171

29. of Thermo Fisher Scientific Inc. see https:// www.thermofisher.com/order/catalog/product/11206D?SID=srch-srp-11206D

30. Mempel TR, Henrickson SE, Von Andrian UH (2004) T-cell priming by dendritic cells in lymph nodes occurs in three distinct phases. Nature 427:154–159

31. Shakhar G, Lindquist RL, Skokos D et al (2005) Stable T cell-dendritic cell interactions precede the development of both tolerance and immunity in vivo. Nat Immunol 6:707–714

Chapter 15

Time-Lapse Förster Resonance Energy Transfer Imaging by Confocal Laser Scanning Microscopy for Analyzing Dynamic Molecular Interactions in the Plasma Membrane of B Cells

Hae Won Sohn and Joseph Brzostowski

Abstract

For decades, various Förster resonance energy transfer (FRET) techniques have been developed to measure the distance between interacting molecules. FRET imaging by the sensitized acceptor emission method has been widely applied to study the dynamical association between two molecules at a nanometer scale in live cells. Here, we provide a detailed protocol for FRET imaging by sensitized emission using a confocal laser scanning microscope to analyze the interaction of the B cell receptor (BCR) with the Lyn-enriched lipid microdomain on the plasma membrane of live cells upon antigen binding, one of the earliest signaling events in BCR-mediated B cell activation.

Key words B lymphocyte, Förster resonance energy transfer (FRET), Confocal laser scanning microscopy, B cell receptor (BCR), Signaling, Lyn

1   Introduction

To fully appreciate how B cells function to provide adaptive immunity in response to the pathogen-driven antigens or autoimmune environments, it is necessary to understand the dynamics of the most initial events in the BCR signaling pathway [1, 2]. B cell activation begins with the clustering of BCRs by antigen-specific binding and crosslinking with a concomitant change in the local lipid microenvironment surrounding the clustered BCRs to allow the association of the BCR with Lyn, the first kinase in BCR-mediated signaling cascade [3, 4]. These initial events occur within a second time scale, making them amenable for FRET measurements by laser scanning confocal microscopy.

Because of the nondestructive nature of FRET detection, the method has been widely applied in living cells to measure the association between two molecules whose respective donor and

Chaohong Liu (ed.), *B Cell Receptor Signaling: Methods and Protocols*, Methods in Molecular Biology, vol. 1707, https://doi.org/10.1007/978-1-4939-7474-0_15, © Springer Science+Business Media, LLC 2018

acceptor fluorescent probes are within a 10 nm distance from one another [5].

FRET is the nonradiative transfer of energy from a donor fluorophore to an acceptor fluorophore by dipole-dipole coupling, resulting in the quenching of donor fluorescent emission and a concomitant increase (sensitization) in acceptor emission. The efficiency of energy transfer is exquisitely sensitive to distance, being inversely proportional to the distance between donor and acceptor fluorophores by a power of 6 [6]. Distance measurements are limited between 1 and 10 nm [7], providing nanometer-scale resolution of molecular interactions that otherwise cannot be measured with conventional fluorescence microscopes. Several techniques have been developed to detect FRET which include acceptor photobleaching, fluorescence lifetime imaging (FLIM) of the donor fluorophore, and sensitized acceptor emission (also called 3^3- FRET) [8, 9]. While the sensitized emission approach requires the most careful controls and corrections of all the methods, it allows FRET measurements to be performed using conventional laser scanning confocal fluorescence microscopy and allows the measurements of spatial and temporal dynamic interactions in live cells (compared to acceptor-photobleaching, an end point method that destroys the acceptor fluorophore, making it useless for dynamic studies, or FLIM, which typically requires slow acquisition times; however, new technological advances in hardware may speed FLIM acquisitions in the near future) [10]. FRET efficiencies using the sensitized emission method can be expressed as either E_d (sensitized emission normalized by the amount of donor) or E_a (sensitized emission normalized by the amount of acceptor) from a set of images captured in three channels: donor (D), FRET (F), and acceptor (A). In a specimen expressing a FRET donor and acceptor pair, the fluorescence signal from the quenched (by energy transfer) donor in D is collected by exciting the specimen with the donor-specific laser and collecting donor emission with a donor-specific emission filter; the signal in F is collected by exciting the specimen with the donor-specific laser and collecting sensitized acceptor emission with an acceptor-specific emission filter; and the signal in A is collected by exciting the specimen with the acceptor-specific laser and collecting acceptor emission with an acceptor-specific emission filter.

FRET efficiency is calculated according to the following equations:

$$E_d = \frac{\mathrm{FR}_d}{K_d + \mathrm{FR}_d} \tag{1}$$

$$E_a = \frac{\mathrm{FR}_a}{K_a} \tag{2}$$

where FR_d and FR_a are determined by

$$\mathrm{FR}_d = \frac{\mathrm{FRET} - \beta \times D - \gamma \times (1 - \beta \times \delta) \times A}{D - \delta \times F} \qquad (3)$$

$$\mathrm{FR}_a = \frac{\mathrm{FRET} - \beta \times D - \gamma \times (1 - \beta \times \delta) \times A}{A \times \gamma \times (1 - \beta \times \delta)} \qquad (4)$$

These equations have been derived previously [9, 10] and modified [11]. D, F, and A are background-subtracted fluorescence intensities in the donor, FRET, and acceptor channels, respectively. β is the correction factor for the donor spectral bleed-through into the FRET channel. β is calculated as F/D, using a donor-only sample excited by the donor-specific excitation laser. γ and δ are correction factors for acceptor crosstalk captured in FRET channel and spectral bleed-through of acceptor emission back into donor channel respectively. γ is calculated as F/A and δ as D/F using an acceptor-only sample excited with both the donor- and acceptor-specific lasers. While the donor in some FRET pairs can be directly excited by the acceptor-specific laser, this cross-talk (which can be corrected for) does not occur with the donor-acceptor pair used in this method and will not be considered. K_d and K_a are constants that convert the amount of sensitized acceptor emission detected in the FRET channel to FRET efficiency that is normalized by donor and acceptor respectively [11]. K_d and K_a are determined from cells expressing a FRET positive control where the donor and acceptor fluorophore are separated with a short amino acid linker at distance close enough to produce nonzero FRET at a 1:1 molar ratio (*see* details below).

Here, we provide a FRET protocol using laser scanning confocal microscopy for live B cells. In this protocol, CH27 B cells stably express Igα-CFP (donor) and Lyn16-YFP (acceptor) to analyze the foremost event of the BCR signaling upon antigen binding to the BCR: the spatial and temporal dynamics of the interaction between the BCR and local lipid microenvironment enriched with Lyn in B cells.

2 Materials

2.1 Stable B Cell Lines and Culture Media

1. Stable cell lines were previously established to express experimental FRET and control constructs and have been explained in detail [12]. All probes localize to the plasma membrane:

    (a) CH27 cells expressing the FRET donor and acceptor pair Igα-CFP and Lyn16-YFP respectively (αCLY).

    (b) CH27 cells expressing only the donor, Igα-CFP chimeric protein (αC).

    (c) CH27 cells expressing only the acceptor, Lyn16-YFP (LY). Lyn16-YFP is an N-terminal fusion of the first

16 amino acid of Lyn to YFP, which targets the construct to the plasma membrane.

(d) Daudi cells expressing Lyn16-CFP-YFP, where CFP and YFP are at a 1:1 molar ratio and separated by 2 amino acids to provide a FRET-positive control (*see* **Notes 1** and **2**).

2. Cell culture medium: Iscove's Modified Dulbecco Medium (IMDM) supplemented with either 15% fetal bovine serum (FBS) for CH27 cells or 10% FBS for Daudi cells, 0.1 mM of GIBCO™ MEM Non-Essential Amino Acids solution, 50 μM β-mercaptoethanol, 100 units/ml of Penicillin and Streptomycin solution (Gibco/BRL, Bethesda, MD, USA).

2.2 Preparation of Chambers and Cells for Imaging

1. Imaging chamber: Lab-TekII Coverglass Chambers with cover, #1.5 Borosilicate, 8 well (Thomas Fisher Scientific Inc., Rochester, NY, USA).

2. Imaging buffer: 1 × Hank's Balanced Salt Solution (HBSS) containing 0.1% FBS (Gibco/BRL). Freshly prepare and filter-sterilize.

3. Cell washing buffer: 1 × HBSS.

4. Deionized ultrapure water: 18.2 MΩ cm resistance and total organic contents less than five parts per billion (ppb) using Milli-Q Quantum Tex filtering system (Millipore).

5. 0.1% (w/v) Poly-L-Lysine stock solution in H_2O: A 0.01% (w/v) Poly-L-Lysine working solution is freshly prepared by diluting the stock solution with deionized ultrapure water.

2.3 FRET Imaging by Confocal Microscopy

1. ZEISS LSM Axiovert 200 M inverted confocal laser scanning microscope equipped with a Plan-Apochromat 63×/1.4 oil immersion objective lens, environmental stage, and lens heater for 37 °C incubation, and 458 nm and 514 nm argon laser to excite CFP and YFP, respectively (Carl Zeiss, Thornwood, NY, USA) (*see* **Note 3**).

2. Soluble antigens for BCR crosslinking: F(ab′)$_2$ goat antibodies specific for mouse IgM (Jackson ImmunoResearch Laboratories, Inc., West Grove, PA, USA).

2.4 Image Processing

1. Image-Pro Plus software (MediaCybernetics, Bethesda, MD).

2. ZEN 2012 (Carl ZEISS).

3. Microsoft Office Excel.

3   Methods

The measurement of FRET by sensitized acceptor emission requires three images D, F, and A and the determination of the correction factors from the control FRET probe cells to subtract bleed-through in the FRET channel from direct D and A excitation. In theory, as long as imaging conditions are identical, control acquisitions need only be performed once; however, in practice, it is best that experimental and control acquisitions are performed the same day to mitigate for changes in microscopic equipment. All the images are processed using pixel-by-pixel arithmetic for FRET efficiency images or calculated for FRET efficiency using Microsoft Excel according to the Eqs. (1–(4)) [11].

3.1   Cell Culture

1. Maintain stable CH27 cell lines expressing *Igα-CFP and/or Lyn16-CFP* (αCLY, αC, and LY) in 15% IMDM culture media in T-25 culture flasks at 37 °C under 7% CO_2 in a humidified incubator. Maintain the stable Daudi cell line expressing *Lyn16-CFP-YFP* (Lyn16-C-Y) in 10% IMDM culture media at 37 °C under 5% CO_2 in a humidified incubator. Passage cells into fresh media when confluent by diluting 1:10 (*see* **Notes 4** and **5**).

2. One day before imaging, dilute cells in fresh culture media at the density of ~0.3×10^6 cells/ml and culture until imaging.

3.2   Preparation of Chamber and Cells for Imaging

1. One day before the imaging, prepare a sterilized 0.01% poly-L-Lysine working solution with sterile ultrapure deionized water. In a biosafety cabinet, place 250 μl of the working solution into each well of 8-well Lab-TekII coverglass chamber (*see* **Note 6**). Incubate for 1 h, completely aspirate the solution, and dry out the chambers in a biosafety cabinet. Maintain sterility until use.

2. Prepare cells for imaging. Transfer the cells that were prepared the previous day to a 15 ml centrifuge tube and pellet at $90 \times g$ at room temperature for 10 min (*see* **Note 7**).

3. Wash the cells twice. Gently resuspend the cell pellet with $1 \times$ HBSS and centrifuge at $90 \times g$. Aspirate the wash buffer and repeat.

4. Resuspend the twice washed cell pellet with $1 \times$ HBSS to a final concentration of 5.0×10^6 cells/ml and transfer 200 μl of cells to the poly-L-lysine-coated Lab-TekII chambers and incubate for 10 min at 37 °C with 5% CO_2.

5. Wash out unattached cells by adding 500 μl of warmed $1 \times$ HBSS buffer to each well and completely remove the buffer with a micropipette. Repeat the wash two more times and finally add 250 μl $1 \times$ HBSS buffer. Maintain cells at 37 °C with 5% CO_2 until imaging.

3.3 Image Acquisition Using a ZEISS LSM Laser Scanning Confocal Microscope

1. At least 1 h prior to the preparation of imaging chambers and cells, turn on the power to the microscope and environmental systems to equilibrate the stage and lens to 37 °C to minimize focal drift.

2. Start the argon gas laser for the 458 and 514 nm lines needed for FRET imaging and allow ~1 h to warm up in standby to minimize power drift.

3. To acquire the three D, F, and A channel images, set up two sequential tracks, one for the simultaneous capture of the donor CFP (D) and FRET (F) signal and the other for the acceptor YFP (A) signal. Select the appropriate excitation laser and emission filters based on what is allowed by your system. For the Zeiss LSM 510, set the first acquisition track as follows:

   CFP (D) = excitation 458 nm, emission BP 475–525 nm.

   FRET (F) = excitation 458 nm, emission LP 530 nm.

   And the second acquisition track as follows:

   YFP (A) = excitation 514 nm, emission LP 530 nm.

4. Set tracks to alternate the acquisition between line scans to minimize the time delay between the sequentially acquired tracks.

5. Specifically for the Zeiss LSM 510, choose the HFT 458/514 main beam splitter to reflect excitation light to the specimen and the NFT 515 dichroic beam splitter to direct the CFP, FRET, and YFP signals to their respective detectors.

6. Optimal imaging modes are system dependent with respect to detector sensitivity and scanning speed. For our live cell imaging we use the following conditions: 63 × 1.40 N.A. oil immersion objective lens; 512 × 512 image size; 12 bit data depth; 1.6 μs pixel dwell time; 4 frame averaging; pinhole size to yield a 3 μm optical section; and a zoom factor of 3 to yield a 0.95 μm × 0.95 μm pixel size.

7. After placing a drop of oil on the lens, insert the Lab-TekII chamber with attached cells and focus using transmitted light or by fluorescence excitation.

8. Scan cells in confocal imaging mode and adjust the PMT detector gain and amplifier offsets to place all fluorescence signals within the 12 bit scale without saturation and with no pixels below zero. Minimize the laser power so as to not photodamage and photobleach cells during time-lapse acquisitions (*see* **Note 8**).

9. Acquire CFP, FRET, and YFP channel images from CFP- or YFP-only expressing cells, αC and LY. Obtain images for ~10 cells with varying fluorescence intensities from low to high signal.

10. Photobleach the acceptor with the FRET-positive construct Lyn16-CFP-YFP to calculate K_d, K_a, $E_{\text{bleaching}}$, FR_d, and FR_a. Acquire CFP, FRET and YFP channel images for the positive FRET cells expressing Lyn16-CFP-YFP (these will be the pre-bleach images for each channel). Next, set a photobleaching routine and completely bleach the YFP signal of the entire cell with the 514 nm laser set at full power. Acquire the CFP, FRET, and YFP channels (post-bleach) with the same acquisition conditions used prior to bleaching. Repeat the process with multiple cells.

11. Next determine the rate of photobleaching in the FRET sample due to image acquisition. Place the chamber containing cells expressing both Igα-CFP and Lyn-YFP (αCLY) onto the environmental stage and allow several minutes for the temperature to equilibrate to maintain focus. Set a time-lapse routine and acquire the D, F, and A channels at 10 s intervals for 10 min under the same imaging conditions of single-probe control cells.

12. Finally, measure the change in FRET efficiency in response to antigen stimulation. Choose another cell area in the same chamber. Start the 10 min time-lapse and acquire 2 scans. Pause the acquisition, add antigen by micropipette to crosslink the BCR and then resume the acquisition for the remaining time.

3.4 Image Processing for the Calculation of FRET and the Creation of FRET Images

1. Open three channel images in the ZEN analysis software (Carl Zeiss) (Fig. 1a) and, if necessary, adjust lateral (x/y alignment) errors using the alignment function found under the Processing Tab. If a flat-field correction is necessary, divide control and FRET images by the normalized reference image and save the corrected images (*see* **Note 9**).

3.4.1 Image Processing for the Calculation of FRET

2. β is the correction factor for the donor (CFP) spectral bleedthrough into the FRET channel. β is determined by exciting cells expressing CFP-only with the 458 nm laser and is calculated as $\beta = \frac{\text{FRET} - F_0}{\text{CFP} - C_0}$ where FRET and CFP are the MFIs obtained from an ROI drawn around the plasma membrane of cell images in the FRET and CFP channel and F_0 and C_0 are the average background intensity in each image obtained from a selected ROI containing no signal from cells. Rearranged FRET $= (F_0 - \beta \times C_0) + \beta \times \text{CFP}$ where $(F_0 - \beta \times C_0)$ is a constant term (β is a constant correction factor and background values ought to remain the same in all images); therefore, the equation then simply describes a line in the form $y = mx + b$, with β as the slope. β is determined experimentally from multiple cells of varying overall intensity by plotting MFI values of FRET and CFP in a scatter plot of FRET vs CFP and

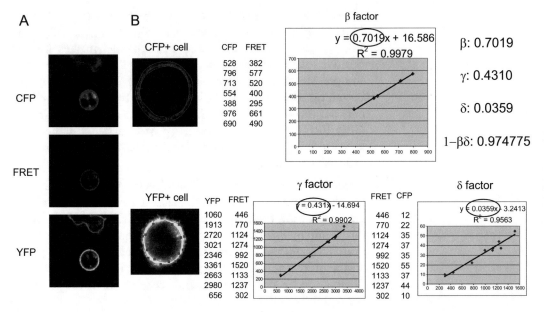

Fig. 1 Three channel images acquired for the FRET imaging by sensitized acceptor emission (**a**) and determination of the correction factors, β, γ, δ (**b**). (**a**) Cells expressing either Igα-CFP alone or Lyn16-YFP alone, or expressing both Igα-CFP and Lyn16-YFP, were settled in the imaging chambers. Cells were excited sequentially with 458 and 514 nm laser light by confocal LSM ZEISS 510 META. Fluorescent emission was simultaneously acquired for the CFP and FRET images (458 excitation) followed by the YFP image (514 excitation) sequentially using line switching mode during the scan and the images were obtained as three channels CFP, FRET, and YFP. (**b**) The correction factors were determined for CFP bleed-through (β), and YFP crosstalk (γ) to the FRET channel by direct CFP or YFP excitation, respectively, and also YFP back-through to the CFP channel (δ) by direct YFP excitation by the 458 nm laser from control cells expressing either Igα-CFP alone (CFP+cell) or Lyn16-YFP alone (YFP+ cell). Regions of interest (ROIs) were drawn over the plasma membrane of cells to obtain mean fluorescence intensity (MFIs) for each channel. To calculate β, the MFI was obtained from ROIs in the CFP and FRET images from multiple cells and was plotted as FRET vs. CFP (*upper graph*). β was determined from the slope of the line in the plot. γ and δ were calculated similarly using the YFP and FRET and FRET and CFP images, respectively, from YFP+ control cells. Representative data and plots are shown

finding the slope obtained from the linear regression (Fig. 1b) (*see* **Note 10**).

3. Likewise, determine γ and δ from cells expressing YFP-only. γ is the correction factor for acceptor (YFP) crosstalk captured in the FRET channel via direct excitation from 458 nm laser and δ is the correction for the spectral bleed-through of the acceptor (YFP) emission back into the CFP channel by 458 nm excitation. $\gamma = \frac{FRET - F_0}{YFP - Y_0}$ where FRET and YFP are the MFI values in the FRET and YFP channel as excited by the 458 and 514 laser, respectively, and F_0 and Y_0 are the intensities of background in the FRET and YFP channels, respectively. $\delta = \frac{CFP - C_0}{FRET - F_0}$ where CFP and FRET are the intensity values in the CFP and FRET channels as excited by the 458 laser and C_0 and F_0 are the intensities of background in each channel. Likewise as done in **step 2**, obtain γ and δ factors from the slope obtained by linear

regression of scatter plots FRET vs. YFP and CFP vs. FRET, respectively (Fig. 1b).

4. Determine the FRET efficiency from acceptor photobleaching, FRET ratios and conversion factors K_a and K_d from images of cells of varying intensity expressing the FRET-positive Lyn16-CFP-YFP fusion protein. These parameters will be used to determine the FRET efficiency in αCLY cells (Fig. 2). FRET efficiency from acceptor photobleaching is

$$E_{\text{bleaching}} = \frac{(\text{CFP}_{\text{after}} - C_0) - \frac{\text{CFP}_{\text{before}} - C_0 - \delta \times (\text{FRET} - F_0)}{1 - \beta \times \delta} \times (1 - \text{Br})}{\text{CFP}_{\text{after}} - C_0 + \frac{\text{CFP}_{\text{before}} - C_0 - \delta \times (\text{FRET} - F_0)}{1 - \beta \times \delta} \times \text{Br}}$$

This equation is the typical efficiency calculation determined from acceptor photobleaching in the form $E = (\text{CFP}_{\text{after}} - \text{CFP}_{\text{before}})/\text{CFP}_{\text{after}}$ now corrected for crosstalk and acquisition photobleaching.

$\frac{\text{CFP}_{\text{before}} - C_0 - \delta \times (\text{FRET} - F_0)}{1 - \beta \times \delta}$ (simplified for equations below as CFPcor) corrects the CFP channel for YFP crosstalk by direct excitation with the 458 nm laser. The laser power used for image acquisition can cause unwanted photobleaching of CFP between scans (acquisition photobleaching) and this signal loss is corrected by the average acquisition photobleaching rate Br. After the YFP acceptor is completely photobleached with the 514 nm laser in cells expressing Lyn16-CFP-YFP, capture three additional scans (i, $i + 1$, $i + 2$). Calculate Br according to this equation $\text{Br} = 1 - \left(\frac{\text{CFP}_{i+1}}{\text{CFP}_i} + \frac{\text{CFP}_{i+2}}{\text{CFP}_{i+1}}\right)/2$.

As in **steps 2** and **3**, the equation can be rewritten in the form $y = mx + b$

$$\text{CFP}_{\text{after}} - \frac{\text{CFP}_{\text{before}} + \delta \times \text{FRET}}{1 - \beta \times \delta} \times (1 - \text{Br}) =$$
$$\left(C_0 - E_{\text{bleaching}} \times C_0 + E_{\text{bleaching}} \times \text{Br} \times \frac{\delta \times F_0 - C_0}{1 - \beta \times \delta} \right.$$
$$+ (1 - \text{Br}) \times \frac{\delta \times F_0 - C_0}{1 - \beta \times \delta}\left. \right)$$
$$+ E_{\text{bleaching}} \times \left(\text{CFP}_{\text{after}} + \frac{\text{CFP}_{\text{before}} + \delta \times \text{FRET}}{1 - \beta \times \delta} \times \text{Br} \right)$$

where

$$C_0 - E_{\text{bleaching}} \times C_0 + E_{\text{bleaching}} \times \text{Br} \times \frac{\delta \times F_0 - C_0}{1 - \beta \times \delta} + (1 - \text{Br})$$
$$\times \frac{\delta \times F_0 - C_0}{1 - \beta \times \delta}$$

is a constant describing the background. The equation is fit with a line as above with the left side (dequenched donor, Υ axis) of the equation and right side, Br-corrected CFP_{after} $\left(CFP_{after} + \frac{CFP_{before} + \delta \times FRET}{1 - \beta \times \delta} \times Br\right)$ (X axis) having the slope of $E_{bleaching}$ (*see* **Note 11**).

The conversion factors between FRET ratios and FRET efficiency are determined from FRET ratios and $E_{bleaching}$:

$$K_a = \frac{FR_a}{E_{bleaching}}$$

where

$$FR_a = \frac{N_{sen}}{(YFP - \Upsilon_0) * \gamma}$$

and

$$K_d = FR_d \times \frac{\left(1 - E_{bleaching}\right)}{E_{bleaching}}$$

$$FR_d = \frac{N_{sen}}{CFP_{cor}}$$

where

$$N_{sen} = \frac{FRET - F_0 - \beta \times (CFP - C_0) - \gamma \times (1 - \beta \times \delta) \times (YFP - \Upsilon_0)}{1 - \beta \times \delta} \quad \text{and}$$

$$CFP_{cor} = \frac{CFP - C_0 - \delta \times (FRET - F_0)}{1 - \beta \times \delta}$$

N_{sen} can be rewritten in a simpler form by substituting CFP_{cor}:

$$N_{sen} = (FRET - F_0) - \beta \times CFP_{cor} - \gamma \times (YFP - \Upsilon_0)$$

Again, the equations are rearranged in the form of $y = mx + b$ so that the FRET ratio can be determined independently of background by the slope.

FR_a is determined from a linear fit to a plot of. $\frac{N_{sen}}{\gamma}$ versus YFP.
FR_d is determined from a linear fit to a plot of. N_{sen} versus CFP_{cor}.

5. FRET efficiency for the donor (E_d) and the acceptor (E_a) from the images of experimental sample cells are calculated as follows:

$$CFP_{cor} = \frac{CFP - \delta \times FRET}{(1 - \beta \times \delta)} \tag{5}$$

$$N_{sen} = FRET - \beta \times CFP_{cor} - \gamma \times YFP \tag{6}$$

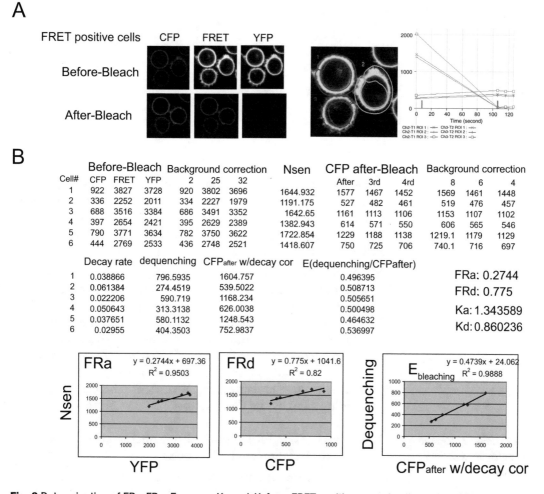

Fig. 2 Determination of FR_d, FR_a, $E_{bleaching}$, K_d, and K_a from FRET positive control cells. (**a**) The CFP, FRET, and YFP images were acquired before and after photobleaching with the 514 nm laser with the same acquisition conditions as described in Fig. 1. Conditions necessary for complete YFP photobleaching were predetermined for cells expressing Lyn16-CFP-YFP by testing different numbers of iterations necessary to eliminate the YFP signal and were used for all subsequent photobleaching experiments (Before-Bleach and After-Bleach images). One pre-bleach and three sequential post-bleach images were obtained and used for the calculation of CFP decay correction. MFI of ROIs over the plasma membrane of cells were obtained for all three channel images in the Zen software (Zeiss). (**b**) The procedure to obtain K_d and K_a from the MFIs of CFP, FRET, and YFP in the Microsoft Excel sheet is shown. Shown are the MFIs after background subtraction before (Before-Bleach) and after bleaching (After-Bleach), Nsen (Eq. 5), CFP Dequenching (CFP$_{after}$ − CFP$_{before}$) by acceptor photobleaching, CFP decay rate, and decay corrected CFP$_{after}$. FR_d and FR_a were determined by the slope of line of N_{sen} vs. CFP and N_{sen} vs. YFP, respectively and $E_{bleaching}$ by the slope of Dequenching vs decay-corrected CFP$_{after}$. Finally K_d and K_a were obtained from the equations, $K_d = FR_d \times (1 - E_{bleaching})/E_{bleaching}$ and $K_a = = FR_a/E_{bleaching}$, respectively

$$FR_d = \frac{N_{sen}}{CFP_{cor}} \qquad (7)$$

$$FR_a = \frac{N_{sen}}{\gamma \times YFP} \qquad (8)$$

$$E_d = \frac{FR_d}{K_d + FR_d} \qquad (1)$$

$$E_a = \frac{FR_a}{K_a} \qquad (2)$$

where the average background value C_0, F_0, and Υ_0 is subtracted from the respective MFI obtained from the ROI drawn around the plasma membrane of cells in the CFP, FRET, and YFP channels using Excel (Fig. 3a). Here, equations 7 and 8 were rewritten form of equations 3 and 4, respectively.

6. For time-lapse FRET images, correct for acquisition photobleaching in the FRET channel. Determine the acquisition photobleaching decay rate using the time-lapse images from unstimulated cells. Plot E_a or E_d (y-axis) as a function of the number of scans (x-axis) and fit the curve with a simple exponential decay $E(n) = E(0) \times e^{-r \times n}$ where $E(n)$ is E_d or E_a after n scans and r determines the rate of decay. For experimental data from stimulated cells, E_d or E_a are then normalized by dividing $E(n)$ by the decay rate, $e^{-r \times n}$ (*see* **Note 12**).

3.4.2 Image Processing for the Creation of FRET Images

1. Save each (CFP, FRET, and YFP) channel as an individual TIFF file from the stimulated cell data set.

2. Open all three channel TIFF files in Image Pro Plus (Media Cybernetics, Rockville, MD, USA) and subtract the average background from the image (Fig. 3b Raw) and, when necessary, smoothen images using a low pass filter (*see* **Note 13**).

3. Determine the threshold value that will select the plasma membrane using the YFP channel image. Using this value, create a masked image to distinguish the plasma membrane from the background. Apply the mask to all three channel images to convert the pixel values of the region outside and inside the cell to zero. This step will remove artificially high FRET in areas driven by random background signal (Fig. 3b masked) (*see* **Note 14**).

4. For time-lapse images, make a separate sequence file for each channel and create a mask for each time point from the YFP channel to be applied as in step 3 to the corresponding time point for all three channels.

5. The final FRET efficiency images are obtained by using the pixel-by-pixel image calculation tool in ImagePro Plus

A

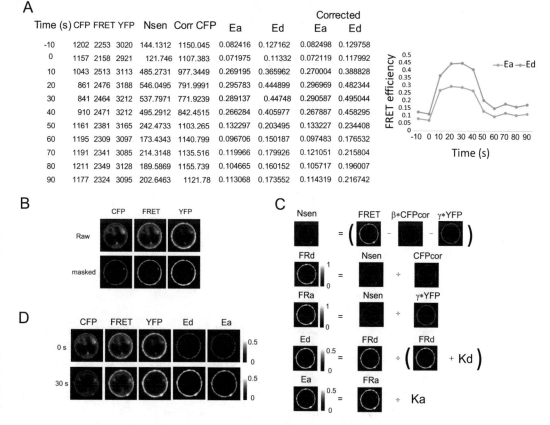

Time (s)	CFP	FRET	YFP	Nsen	Corr CFP	Ea	Ed	Corrected Ea	Corrected Ed
-10	1202	2253	3020	144.1312	1150.045	0.082416	0.127162	0.082498	0.129758
0	1157	2158	2921	121.746	1107.383	0.071975	0.11332	0.072119	0.117992
10	1043	2513	3113	485.2731	977.3449	0.269195	0.365962	0.270004	0.388828
20	861	2476	3188	546.0495	791.9991	0.295783	0.444899	0.296969	0.482344
30	841	2464	3212	537.7971	771.9239	0.289137	0.44748	0.290587	0.495044
40	910	2471	3212	495.2912	842.4515	0.266284	0.405977	0.267887	0.458295
50	1161	2381	3165	242.4733	1103.265	0.132297	0.203495	0.133227	0.234408
60	1195	2309	3097	173.4343	1140.799	0.096706	0.150187	0.097483	0.176532
70	1191	2341	3085	214.3148	1135.516	0.119966	0.179926	0.121051	0.215804
80	1211	2349	3128	189.5869	1155.739	0.104665	0.160152	0.105717	0.196007
90	1177	2324	3095	202.6463	1121.78	0.113068	0.173552	0.114319	0.216742

B

C

D

Fig. 3 Calculation of FRET efficiencies, E_d and E_a and processing to make FRET efficiency images from time-lapse confocal LSM FRET imaging in live B cells. Cells were settled and time-lapse FRET images were acquired at 37 °C in 10 s interval for 10 min. (**a**) The procedure to obtain the bleach rate-corrected E_d and E_a from the MFIs of CFP, FRET, and YFP for time-lapse FRET imaging in the Microsoft Excel sheet is shown. Shown are the MFIs after background subtraction of ROIs drawn over the plasma membrane of cells at each time point for each channel, N_{sen}, and CFP$_{cor}$ as calculated in Fig. 3b. E_a and E_d were calculated and the acquisition bleach-rate obtained from the control scan from the experimental cells was applied to E_d and E_a resulting in the calculation of acquisition photobleach-corrected E_d and E_a. (**b, c**) The procedure for image processing to obtain the FRET efficiencies, E_d and E_a images from experimental cells expressing both Igα-CFP and Lyn16-YFP is shown. Three channel images taken before and after cell stimulation with antigen were first background subtracted (**b**, Raw) and masked (**b**, masked) for plasma membrane using YFP images as described in the text. (**c**) $\beta \times$ CFP$_{cor}$, $\gamma \times$ YFP and $\delta \times$ FRET images were made by multiplying β, γ, and δ values with the CFP$_{cor}$, YFP, and FRET images, respectively. CFP$_{cor}$, N_{sen}, FR$_a$, FR$_d$, E_d, and E_a images were obtained by the arithmetic image calculation following the Eqs. (1, 2, 5–8) described in the text. Shown are the 12 bit gray scale images for CFP, FRET, YFP, $\beta \times$ CFP$_{cor}$, $\gamma \times$ YFP, CFP$_{cor}$, and N_{sen} images. For the ratiometric images FR$_d$, FR$_a$, E_d, and E_a, an arbitrary scale factor was multiplied to the final images and smoothened by low pass filtering for the proper display. For each image, a relative intensity scale bar is shown. (**d**) CFP, FRET, YFP, E_d, and E_a images before (0 s) or after stimulation (30 s) with soluble antigen are shown. Shown are the raw CFP, FRET, and YFP images and the masked E_d and E_a images at 0 s and 30 s. E_d and E_a are shown as gray scale images from 0 to 0.5. The CFP, FRET, and YFP images are 12bit gray scale

programed with the equations below. Use the background subtracted and masked images for all calculations (Fig. 3c):

$$CFP_{cor} = \frac{CFP - \delta \times FRET}{(1 - \beta \times \delta)} \tag{5}$$

$$N_{sen} = FRET - \beta \times CFP_{cor} - \gamma \times YFP \tag{6}$$

$$FR_d = \frac{N_{sen}}{CFP_{cor}} \tag{7}$$

$$FR_a = \frac{N_{sen}}{\gamma \times YFP} \tag{8}$$

$$E_d = \frac{FR_d}{K_d + FR_d} \tag{1}$$

$$E_a = \frac{FR_a}{K_a} \tag{2}$$

(*see* **Note 15**)

6. First, produce the corrected CFP (CFP_{cor}) image from Eq. (5). The $\delta \times FRET$ image is obtained by multiplying the masked FRET image by the previously determined correction factor δ. Next, using the pixel-by-pixel image calculation tool, subtract $\delta \times FRET$ from CFP. Divide the resulting image by $1 - \beta \times \delta$.

7. Next, make the sensitized acceptor emission (N_{sen}) image according to the Eq. (6) using the masked CFP, FRET and YFP images in a similar way of processing in **step 6**.

8. Process further to make FR_d and FR_a according to the Eqs. (7) and (8), respectively, using the N_{sen} and CFP_{cor} images for FR_d and N_{sen} and YFP for FR_a (*see* **Note 16**).

9. Create final FRET efficiency images E_d using the FR_d image and K_d, and E_a using FR_a and K_a (*see* **Note 17**) (Fig. 3d).

4   Notes

1. CH27 is a mouse B cell lymphoma cell line that expresses a cell surface IgM BCR that recognizes phosphorylcholine (PC). Daudi is a human Burkett lymphoma cell line that also expresses cell surface IgM BCR (purchased at ATCC). Generally, B cell lines are difficult to stably transfect with the exception of a few well-known cell lines such as mouse A20 and human Daudi cells. Depending on the purpose of study, a transient transfection may be used.

2. Igα was fused in-frame to the N-terminal coding sequence of CFP using pECFP-N1 vector (BD Biosciences Clontech, Palo Alto, CA) at the Bam HI site upstream of the CFP, resulting in a six amino acid spacer between Igα and CFP. While we found that this spacer length gave us the best expression of this fusion proteins at the plasma membrane, other designs may need to consider for alternative constructs. Full length Igα (Cat. No. MMM1013-7510126) is available (Open Biosystems, Huntsville, AL) for PCR templates. The membrane-targeting plasmids, either Lyn16-CFP-YFP or Lyn16-YFP, have been previously described [12].

3. The 458 nm laser used in this protocol is only ~80% efficient for CFP excitation and directly activates the YFP acceptor with ~40% efficiency, requiring significant correction for crosstalk. CFP excitation can be maximized and YFP crosstalk minimized by using a system equipped with a shorter wavelength diode laser.

4. CH27 cells grow better in 7% CO_2 than the typical 5% mixture. It is important to grow CH27 at 7% CO_2 when first thawed from liquid nitrogen; however, once established they can be cultured at either condition. Cells are usually passaged every 3 days. Keep track of cell density by counting. When cells reach confluency at about 1.5×10^6 cells/ml, dilute cells 1:10 into fresh media.

5. Although transient transfections may be less time-consuming, we found that creating stable cell lines with a proper ratio (~1:1) of CFP to YFP worked best in FRET experiments. The ratio of CFP and YFP expression in cells can be determined by comparing the CFP and YFP MFI of test cells to those of cells expressing a CFP-YFP fusion protein with spacer far enough apart to provide zero-FRET. In this case, we used cells expressing the CFP-TRAF2DN-YFP chimeric protein with a structurally restricted 232 amino acid spacer between CFP and YFP [13]. The retroviral spin-infection system using the MSCV vector and the PT67 packaging cell line (Clontech) followed by FACS sorting for positives [14] works best for CH27 cells. Stable cell lines expressing both CFP and YFP used here were established by sequential or simultaneous infection followed by repetitive sorting for CFP and/or YFP positive cells. We performed "spin infection" [14] by centrifugation at 1800 × g for 90 min followed by the concentration of the viral supernatant by centrifugation at 3000 × g for about 20 min using a centrifugal filter device (Amicon Ultra-15 with 10k MW cut-off, Cat No. UFC901008, Millipore). Stable Daudi cell lines for the FRET positive control were established by drug selection and then sorted by FACS after electroporation according to the manufacturer's protocol using the Bio-Rad gene pulser system.

6. It is critical to attach cells to limit motion for time-lapse image acquisition and subsequent fluorescence intensity measurements at the plasma membrane to determine the FRET signal. Poly-L-Lysine coating allows B cells to adhere to the coverglass without causing stimulation.

7. For cells to tightly attach to poly-L-lysine-coated chambers, proteins should be removed from the growth media by washing with $1 \times$ HBSS (or PBS) and resuspended in the same buffer.

8. For live cell imaging, minimizing laser power is essential to lessen photo-toxic effects. As a consequence, higher detector gains are used to detect signal with averaging to lessen the electronic noise that results from elevated gain. For the LSM 510 META, the PMT gain was set to 800 V with ~9% 458 nm and ~1% 514 nm laser power.

9. If necessary, check uneven illumination across the imaging plane by obtaining a normalized reference image using FITC dye. To obtain a reference image, place 250 μl of diluted dye into a well of a Lab-TekII chamber at a concentration so that no pixels are saturated. Collect a raw reference image using the same imaging conditions as used for FRET imaging: the same number of pixels, dwell time, averaging, objective lens, and zoom factor. The normalized reference image is obtained by dividing each pixel value of the raw reference image by the mean intensity of the entire image. If a flat-field correction is required for the FRET images, each channel is divided by the normalized reference image. Pixel shift in the lateral direction can occur on confocal laser scanning systems due to misaligned pinholes or collimating lenses. Lateral alignment is critical in FRET measurements where the region of interest spans only a small number of pixels such as the plasma membrane. Check that pixels in the D, F, and A channels are aligned using beads that fluoresce in all three channels. Make the necessary adjustment to the hardware. If pixel shift persists, align data using software adjustments based on what is determined with control beads.

10. Alternatively, MFI values taken from a line profile drawn over the plasma membrane instead of an ROI.

11. With some exceptions, such as degradation of the fusion protein in certain cells, FRET positive control cells empirically show about $55 \pm 10\%$ $E_{\text{bleaching}}$ regardless of imaging conditions.

12. Acquiring a sequential time-lapse series requires minor corrections for acquisition photobleaching-induced decrease in FRET. In general, E_a drops with the photobleaching of CFP and E_d drops with the photobleaching of YFP. To determine the decay rate of the FRET signal, a resting cell expressing the

same combination of donor and acceptor is imaged for the desired number of sequential scans (there is a basal level of FRET in unstimulated cells). Finally, E_a usually has lower acquisition photobleaching decay in time-lapse imaging because photobleaching of CFP under high FRET conditions is lower than YFP.

13. In this protocol, smoothen images is optional because images recorded are from averaged pixel values of four line scans; thus, low pass filtering does not make significant changes to the quality of the FRET images. However, if images are acquired without averaging, a low pass filter can reduce the noise of the individual pixels.

14. Using a mask function and creating masked images before producing FRET images can remove artificial positive FRET signals derived from the random background signal. As a rule of thumb for determining an appropriate threshold, use the mean plus three standard deviations of the pixel intensity of the background.

15. If the δ factor is negligible by using a shorter wavelength donor excitation lasers, N_{sen} can be simplified as follows:

$$N_{sen} = FRET - F_0 - \beta \times (CFP - C_0) - \gamma \times (YFP - \Upsilon_0)$$

16. For the display purposes, especially for ratiometric images such as FR_d, FR_a, E_d, and E_a, a scale factor should be applied to all pixels, which is an arbitrary number to make possible for the low pixel value images to be seen within range after calculation. 1000 is usually multiplied to each pixel for this purpose. It is important to use an intensity scale bar in the FRET images for comparison purposes.

17. To make blur images for display purposes, a low pass filter may apply to the final FRET efficiency images and also the gray scale images may be changed into color-coded images to show the pixel-by-pixel FRET efficiency for better distinction.

Acknowledgments

This work was supported by the Intramural Research Program of the National Institute of Allergy and Infectious Diseases, National Institutes of Health.

References

1. Fearon DT, Carroll MC (2000) Regulation of B lymphocyte responses to foreign and self-antigens by the CD19/CD21 complex. Annu Rev Immunol 18:393–422

2. Dal Porto JM, Gauld SB, Merrell KT, Mills D, Pugh-Bernard AE, Cambier J (2004) B cell antigen receptor signaling 101. Mol Immunol 41:599–613

3. Pierce SK (2002) Lipid rafts and B-cell activation. Nat Rev Immunol 2:96–105

4. Tolar P, Sohn HW, Pierce SK (2008) Viewing the antigen-induced initiation of B-cell activation in living cells. Immunol Rev 221:64–76

5. Ciruela F (2008) Fluorescence-based methods in the study of protein-protein interactions in living cells. Curr Opin Biotechnol 19:338–343

6. Forster T (1948) Intermolecular energy migration and fluorescence. Ann Phys 2:55–75

7. Truong K, Ikura M (2001) The use of FRET imaging microscopy to detect protein-protein interactions and protein conformational changes in vivo. Curr Opin Struct Biol 11:573–578

8. Butz ES, Ben-Johny M, Shen M, Yang PS, Sang L, Biel M, Yue DT, Wahl-Schott C (2016) Quantifying macromolecular interactions in living cells using FRET two-hybrid assays. Nat Protoc 11:2470–2498

9. van Rheenen J, Langeslag M, Jalink K (2004) Correcting confocal acquisition to optimize imaging of fluorescence resonance energy transfer by sensitized emission. Biophys J 86:2517–2529

10. Zal T, Gascoigne NR (2004) Photobleaching-corrected FRET efficiency imaging of live cells. Biophys J 86:3923–3939

11. Tolar P, Sohn HW, Pierce SK (2005) The initiation of antigen-induced B cell antigen receptor signaling viewed in living cells by fluorescence resonance energy transfer. Nat Immunol 6:1168–1176

12. Sohn HW, Tolar P, Jin T, Pierce SK (2006) Fluorescence resonance energy transfer in living cells reveals dynamic membrane changes in the initiation of B cell signaling. Proc Natl Acad Sci U S A 103:8143–8148

13. He L, Olson DP, Wu X, Karpova TS, JG MN, Lipsky PE (2003) A flow cytometric method to detect protein-protein interaction in living cells by directly visualizing donor fluorophore quenching during CFP→YFP fluorescence resonance energy transfer (FRET). Cytometry A 55:71–85

14. Richie II, Ebert PJ, Wu IC, Krummel MF, Owen JJ, Davis MM (2002) Imaging synapse formation during thymocyte selection: inability of CD3ζ to form a stable central accumulation during negative selection. Immunity 16:595–606

Chapter 16

Understanding of B Cell Receptor Signaling Through a Photo-Activatable Antigen Presentation System

Jing Wang, Zhengpeng Wan, and Wanli Liu

Abstract

Antibody responses are initiated by the binding of antigens to cell surface expressed B cell receptors (BCRs) that trigger signaling cascades resulting in the activation of B cells. However, it has been difficult to study these cascades due to their fast, dynamic, and transient nature. Using a conventional antigen-presenting system, such as planar lipid bilayers (PLBs), the initial events during B cell activation have been difficult to be captured. Here, we describe the general procedures for the utilization of a photoactivatable antigen-presenting system in the combination with a high speed live cell imaging method to investigate the early activation events in the same B cells in response to antigen stimulation.

Key words BCR activation, Photoactivatable antigen, B cell receptor

1 Introduction

B lymphocytes are responsible for antibody responses arising from the recognition of pathological antigens by the cell membrane expressed B cell receptor (BCR) [1]. The BCR is composed of a membrane-bound immunoglobulin (mIg) and a non-covalently associated heterodimer of Igα and Igβ [2, 3]. Antigen binding induces a series of dynamic changes and transformations in both biophysical behaviors and biochemical features of the BCRs, which further determine the fate of a cell [4–6]. However, it has been difficult to accurately capture and comprehensively investigate these changes because of the instantaneity and rapidity of responses after antigen recognition [4, 7]. For example, recent live cell imaging studies showed that the B lymphocytes swiftly accumulate BCRs into the contact interface between the B cells and the antigen-presenting substrates to form a specialized membrane structure, the B cell immunological synapse (IS) in a short time [4, 7]. What is more, studies from ours and others showed that the initiation of B cell activation can be regulated by antigen density [8, 9], affinity [8, 9], valency [10–12], mobility [13], substrate stiffness [14, 15],

Chaohong Liu (ed.), *B Cell Receptor Signaling: Methods and Protocols*, Methods in Molecular Biology, vol. 1707,
https://doi.org/10.1007/978-1-4939-7474-0_16, © Springer Science+Business Media, LLC 2018

and mechanical forces [16, 17], implying the early events for B cell activation since BCRs that recognize antigens are organized in a sophisticated way, which is the fundamental mechanism for B cells to response to antigens. However, it is technically difficult to accurately capture these events due to their fast and dynamic nature. It is challenging to capture an entire molecular event from the same B cell before and immediately after antigen recognition. In our previous studies [18], we developed a precisely controllable antigen-presenting system by using photoactivatable antigens, which are in an inactive form unless exposed to UV light. The photoactivation process is quite fast and sensitive, since caged antigens will become immediately activated upon the illumination of UV light. We caged the widely used model hapten antigen 4-hydroxy-3-nitrophenyl acetyl (caged-NP), which after photoactivation can be recognized by NP-specific B1-8-BCR-expressing B cells. The NP hapten is specifically recognized by B1-8 antibody or B1-8-BCR-expressing B cells mainly due to the hydroxyl ($-$OH) group in the phenol ring [19]. By conjugating a UV sensitive moiety 4, 5-Dimethoxy-2-nitrobenzyl (DMNB) to its $-$OH group to generate DMNB-NP, we can mask the antigenicity of NP (Fig. 1a). Upon photoactivation by UV photons, the caging group can be photolysis and then NP hapten can be exposed and recognized by B1-8 antibody or B1-8-BCR-expressing B cells. For more details please refer to our published paper [18]. Taking advantage of this photoactivatable antigen-presenting system, we are able to provide an accurately controllable trigger system to perform high-resolution temporal analyses of the early events of B cell activation.

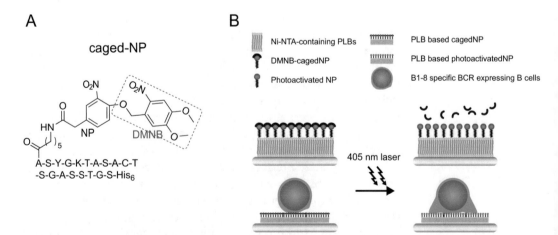

Fig. 1 Caged-NP peptide and TIRFM imaging system setup. (**a**) The schematic description of the UV sensitive moiety, DMNB (caging group), conjugated to the $-$OH group of the NP hapten to generate non-antigenicity DMNB-NP. (**b**) The schematic presentation of the Ni-NTA-containing planer lipid bilayer-based caged-NP present system

In this chapter, we describe how to prepare planar lipid bilayers (PLBs)-based photoactivatable antigen-presenting system and B cells for imaging by total internal reflection fluorescence microscopy (TIRFM), provide a detailed description of the usage of necessary components to set up the lipid bilayer photoactivatable antigen-presenting system (caged-NP), and detail live cell imaging protocols used for the photoactivation of caged-NP antigens.

2 Materials

Prepare phosphate buffer saline (PBS) for the required experimental steps. Prepare all the solutions and reagents using ultrapure water and analytical grade regents. All the reagents should be stored at room temperature (unless indicated otherwise).

2.1 Mice, Cells, Antibodies

1. NP-specific J558L-B1-8-IgM B cells [8].
2. Purify B1-8 primary B cells from the spleen of $IgH^{B1-8/B1-8}$ $Ig\kappa^{-/-}$ transgenic mice [8].
3. Cy5-conjugated Fab anti-mouse IgM constant region antibodies (see **Note 1**).

2.2 Chamber Preparation

1. For live cell imaging, glass chamber slices are generally used (VWR micro cover glass, No. 1.5).
2. 8-well Lab-Tek chamber (see **Note 2**).
3. Rinsing solution: 100% EtOH.
4. Piranha solution: 70% H_2SO_4 add 30% H_2O_2 (see **Note 3**).
5. EZ-spread plating glass beads (MP Biomedicals).
6. Deionized ultrapure water, which has a resistivity of 18.2 MΩ-cm and total organic contents less than 5 parts per billion (ppb).
7. Sylgard 164 Silicone Elastomer adhesive (Dow Corning Corp).
8. Imaging buffer: PBS containing 0.1% FBS (Gibco).
9. Ni-NTA-containing small uniamellar vesicles (SUVs) solution in PBS: 0.1 mM of 1,2-dioleoyl-sn-glycero-3-phosphocholine (DOPC) (Avanti Polar Lipids), 0.04 mM of 1,2-dioleoyl-sn-glycero-3-[N(5-amino-1-carboxypentyl) imino-diacetic acid]-succinyl (Nickel Salt) (DOGS-Ni-NTA; Avanti Polar Lipids) in PBS [8].
10. Blocking buffer: weigh 10 mg casein in 800 mL PBS buffer, 56 °C heated and stirred for some time. Once the solid powder is dissolved, add PBS to 1000 mL. Before storing at −20 °C, filtrate by 0.22 μm ultra-filtrate membrane.

2.3 TIRFM

1. Optical table components and design (LSXPT).

2. Olympus IX-81 microscope.

3. ANDOR iXon + DU-897D electron-multiplying EMCCD camera.

4. 405 nm, 100 mW, argon gas laser (Coherent); 647 nm, 80 mW krypton/argon gas laser ~ 20 mW each line (Coherent).

5. Filter wheel and control box (Applied Scientific Instrumentation).

6. AOTF and control box (NEOS Technologies).

7. Fiber optic launch (OZ Optics).

8. Objective lenses: 100×/1.49 NA UAPON100XOTIRF (Olympus).

9. Focus control (*see* **Note 4**).

10. Temperature control and CO_2 culture environment (*see* **Note 5**).

11. Metamorph system.

3 Methods

3.1 Caged-NP Presenting Preparation

1. Place 24 × 50 mm #1.5 cover glass into Piranha solution overnight to clean the slice. Take the glass out and soak into a deionized ultrapure water filled jar, wash with deionized ultrapure water for 100 mL. Then use 100% EtOH rinse several times for 10 mL. Blow dry completely with argon gas. Place the cover glass on top of pre-cleaned glass beads.

2. Tear off the bottom cover glass of the 8-well Lab-Tek chamber and replace with the cover glass from **step 1** using Sylgard 164. Allowing 10–20 min for adhesion and drying (*see* **Note 6**).

3. Load 200 μL Ni-NTA-containing SUVs solution in PBS into each of the chamber wells, wait for 30 min at 37 °C. Carefully rinse the PLBs with about 10 mL of PBS per well, keeping the bilayers under the solution at all times. Left 150–200 μL PBS in the chamber after the last rinse (*see* **Note 7**).

4. For tethering His_6-tag contained caged-NP to the PLBs, load 200 μL of caged-NP in PBS at the concentration of 10 μg/mL, incubation for 30 min at 37 °C, and wash unbound caged-NP as mentioned above (*see* **Note 8**).

5. Add 200 μL of blocking buffer for 1 h at 37 °C.

6. Before the final washing, label B cell surface receptors BCR. Generally 1×10^6 J558L-B1-8-IgM B cells or B1-8 primary B cells should be labeled with Cy5-conjugated Fab anti-mouse IgM constant region antibodies at 100–200 nM in 200 mL PBS for 10 min at 4 °C, followed by washing twice in PBS (*see* **Note 9**).

7. Rinse the casein blocked chamber with about 10 mL of PBS per well. Immediately before imaging, exchange PBS with imaging buffer (*see* **Note 10**).

8. The schematic in Fig. 1b demonstrates the caged-NP presenting system.

3.2 Photoactivation of Caged-NP and B Cell Receptor Imaging by TIRFM

The behavior of NP-specific B cells before and after photoactivation of caged-NP was examined by TIRFM imaging, the photoactivation experiments by the 405 nm laser can be performed. In this method, we used the Olympus IX-81 microscope equipped with a TIRF port, ANDOR iXon + DU-897D electron-multiplying EMCCD camera, Olympus 100 × 1.49 NA objective TIRF lens, a 405 nm, and a 647 nm laser (Coherent). The acquisition was controlled by the Metamorph system (MDS Analytical Technologies). All live cell imaging experiments were performed at 37 °C. More information about how to use TIRFM and others can be found within the following references: [20–24].

1. Cells are washed three times in PBS after labeling, resuspended in 200 μL imaging buffer. Then load cells to PLBs presented caged-NP chamber wells. After 10 min incubation time, set TIRFM software and focus cells to image plane (*see* **Note 11**).

2. First you can image the basal behaviors of a single B cell in its quiescent state on cover glass presenting the caged-NP for a sufficient amount of time (e.g., 10 min) and then examine the behavioral changes of the same B cell immediately upon photo-activation and thereafter for another 10 min or longer as needed (*see* **Note 12**).

3. The acquired TIRF image movies are analyzed and processed using a combination of Image Pro Plus (Media Cybernetics) and Matlab (Mathworks) software as described before [18] (Fig. 2a–c, movie 1) (*see* **Note 13**).

3.3 Single-Molecular Tracking in Caged-NP System

In order to detect the changes in the BCR before and immediately after NP binding that result in oligomerization and microcluster formation, we tracked single molecule of BCRs through TIRFM. More details can be found in published paper from Tolar's lab [25].

1. Cells are labeled for single-molecule TIRF imaging by incubation 1×10^6 cells with fluorescently labeled BCR-specific Fab fragments at subnanomolar concentration (20–300 pM) in 200 μL PBS for 10 min at 4 °C (*see* **Note 14**).

2. After wash for three times in PBS, resuspended in 200 μL imaging buffer, load cells to PLBs presented caged-NP chamber wells. After 10 min incubation time, set TIRFM software and focus cells to image plane. Find the target cells with 10 mW of laser (at the objective lens in epifluorescence to find the cells and then change to the TIRF angle).

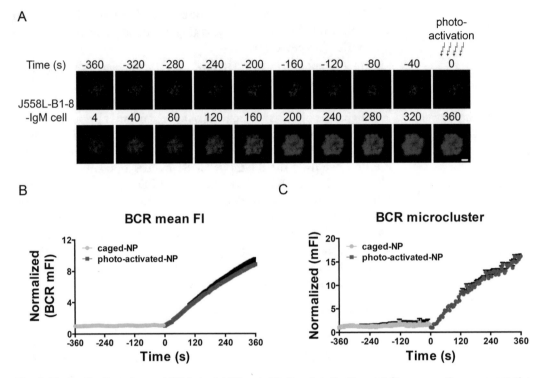

Fig. 2 Photoactivation of caged-NP induced NP-specific B cell activation. (**a**) Shown are the representative TIRFM images at the indicated time points of the same J558L-B1-8-IgM B cells that were placed on cover glass coated with caged-NP reaction before and immediately after photoactivation. (**b, c**) Statistical comparison to quantify the mFI of the BCRs and BCR microclusters mFI from the same B cells before and after photoactivation as depicted in (**a**)

3. To acquire images of single-BCR molecules, an around 100 × 100 pixel region is required.

4. Each B cell was subjected to single-molecule imaging (SMI) of 600 frames in 18 s (30 ms/frame) for the first SMI reading. The second SMI reading was taken after an interval of 10 s followed by the third SMI reading and another 10 s interval until the sixth reading. Photoactivation by 405 nm laser was only executed during the first and second reading. The schematic (Fig. 3) depicts our operation strategy for the single-molecule imaging (SMI) experiment of BCRs (*see* **Note 15**).

4   Notes

1. You can use different fluorophore-conjugated Fab anti-mouse IgM constant region antibodies to label your cells except 405 nm laser excited fluorescent proteins.

2. Here we used 24 × 50 mm No. 1.5 cover glass and 8-well Lab-Tek chamber.

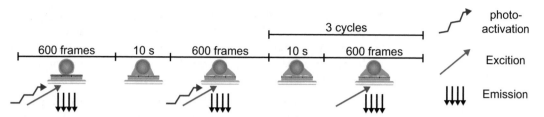

Fig. 3 The schematic representation indicates the experimental strategy during the single-molecule imaging (SMI) of the BCRs. B cell was subjected to single-BCR molecule imaging of 600 frames in 18 s (30 ms/frame) for the first SMI reading. The second SMI reading was taken after an interval of 10 s followed by the third SMI reading and another 10 s interval until the sixth reading. Photoactivation was only applied during the first and second TIRFM imaging cycles

3. Starting with 30% hydrogen peroxide and 98% sulfuric acid, the mixing ratio should be in a range of 3:7 (H_2O_2: H_2SO_4). This solution is very dangerous when preparing this solution, you need special protection equipment including: a full face shield, heavy duty rubber gloves (regular Nitrile gloves will not provide sufficient protection), as well as an acid apron to wear on top of the lab coat. The protective wear should be carefully stored to avoid potential contamination. Slowly add the correct volume of H_2O_2 into H_2SO_4 containing jar, whenever handling Piranha, only use glass containers and always add the peroxide to the acid very slowly. Piranha solution is very energetic and potentially explosive, which is very likely to become hot, even more than 100 °C. Handle with care. Picking up a beaker that is this hot will be very painful might melt your gloves, and may cause you to spill it! Leave the hot piranha solution in an open container until cool.

4. Because the focal plane in TIRFM is extremely narrow (~100 nm), focus control becomes a key issue for time-lapse acquisitions.

5. Additional challenges are faced if the live specimen must be maintained above ambient temperature and within CO_2 culture environment during imaging. The specimen chamber can be heated and CO_2 exchange directly in a closed-chamber system.

6. Prepare chambers on the day of use. Avoid any contamination either on the cover glass or the chamber.

7. It is very important to maintain PLBs under hydrated conditions to avoid any oxidation-induced modifications to PLBs [26]. When washing the chamber, gentle wash is needed.

8. Caged-NP peptide was dissolved by analytically pure DMSO to 0.1 mg/mL. Aliquot to several tubes is highly recommended, keep them in −20 °C, avoid direct light.

9. It is important to use antibody Fabs for labeling cell surface proteins to avoid receptor crosslinking, internalization, and inadvertent engagement of Fc receptors. The fluorophore-conjugated Fabs of specific Abs can be purchased from a number of vendors.

10. PLBs are less stable in imaging buffer, so after finish preparing caged-NP presenting chamber, use within 1–2 h if possible. Caged-NP is very sensitive to light especially UV light, so all the preparing step must avoid being direct exposure to light.

11. Cells were first placed on cover glass presenting the caged-NP antigen for 10 min to blunt any potential behavioral changes of the B cells that were introduced into the system by the acute landing and adhesion responses of the B cells.

12. The acquisition is controlled by Metamorph software using its multiple dimensional acquisition mode (Molecular Devices), with a 1 s exposure time for 405 nm laser and 100 ms for 647 nm laser images, the interval time between each cycle was adjusted to 4 s automatically. It is also important to consider minimizing bleaching effects by titrating the output laser power (here we use 405 nm laser intensity 20%, 647 nm laser intensity 2%). All the photoactivation and TIRFM imaging were performed at 37 °C. The total photoactivation time of 3 s was achieved by 3 cycles of the following imaging acquisition itinerary. Starting from the 4th cycle, the 405 nm photoactivation step was removed from the itinerary, keep 647 nm for BCR imaging.

13. In our case, only the first 360 s of each track of the BCR microclusters from J558L-B1-8-IgM B cell or B1-8 primary B cell microclusters are selected for full analyses. This is necessary to avoid microcluster tracking and 2D Gaussian fitting errors. For the FI values of each BCR microcluster trajectory, values belonging to the same track are normalized to the first position of each track [27, 28]. Or it is available through the online server at http://bioinf.wehi.edu.au/software/com pareCurves/index.html.

14. The concentration of labeling antibody should be low enough that single-BCR molecules can be visualized without the need of photo-bleaching [25]. Single molecules show a single-step fluorescence bleaching profile, providing a diagnostic test for the single-molecule images.

15. Single-molecule tracking of BCR molecules was analyzed as described in our pervious study [18]. MSD and shot-range diffusion coefficients for each BCR molecule trajectory were calculated from positional coordinates and plotted as CDP.

Acknowledgments

This work is supported by funds from Ministry of Science and Technology of China (2014CB542500) and National Science Foundation China (81422020 and 81621002).

References

1. Kurosaki T, Shinohara H, Baba Y (2010) B cell signaling and fate decision. Annu Rev Immunol 28:21–55

2. Schamel WW, Reth M (2000) Monomeric and oligomeric complexes of the B cell antigen receptor. Immunity 13(1):5–14

3. Tolar P, Sohn HW, Pierce SK (2005) The initiation of antigen-induced B cell antigen receptor signaling viewed in living cells by fluorescence resonance energy transfer. Nat Immunol 6(11):1168–1176

4. Pierce SK, Liu W (2010) The tipping points in the initiation of B cell signalling: how small changes make big differences. Nat Rev Immunol 10(11):767–777

5. Dustin ML, Groves JT (2012) Receptor signaling clusters in the immune synapse. Annu Rev Biophys 41:543–556

6. Davis MM et al (2007) T cells as a self-referential, sensory organ. Annu Rev Immunol 25:681–695

7. Harwood NE, Batista FD (2010) Early events in B cell activation. Annu Rev Immunol 28:185–210

8. Liu W, Meckel T, Tolar P, Sohn HW, Pierce SK (2010) Antigen affinity discrimination is an intrinsic function of the B cell receptor. J Exp Med 207(5):1095–1111

9. Fleire SJ et al (2006) B cell ligand discrimination through a spreading and contraction response. Science 312(5774):738–741

10. Bachmann MF et al (1993) The influence of antigen organization on B cell responsiveness. Science 262(5138):1448–1451

11. Liu W, Chen YH (2005) High epitope density in a single protein molecule significantly enhances antigenicity as well as immunogenicity: a novel strategy for modern vaccine development and a preliminary investigation about B cell discrimination of monomeric proteins. Eur J Immunol 35(2):505–514

12. Liu W et al (2004) High epitope density in a single recombinant protein molecule of the extracellular domain of influenza a virus M2 protein significantly enhances protective immunity. Vaccine 23(3):366–371

13. Wan Z, Liu W (2012) The growth of B cell receptor microcluster is a universal response of B cells encountering antigens with different motion features. Protein Cell 3(7):545–558

14. Wan Z et al (2013) B cell activation is regulated by the stiffness properties of the substrate presenting the antigens. J Immunol 190(9):4661–4675

15. Zeng Y et al (2015) Substrate stiffness regulates B-cell activation, proliferation, class switch, and T-cell-independent antibody responses in vivo. Eur J Immunol 45(6):1621–1634

16. Natkanski E et al (2013) B cells use mechanical energy to discriminate antigen affinities. Science 340(6140):1587–1590

17. Wan Z et al (2015) The activation of IgM- or isotype-switched IgG- and IgE-BCR exhibits distinct mechanical force sensitivity and threshold. eLife 4

18. Wang J et al (2016) Utilization of a photoactivatable antigen system to examine B-cell probing termination and the B-cell receptor sorting mechanisms during B-cell activation. Proc Natl Acad Sci U S A 113(5):E558–E567

19. Sonoda E et al (1997) B cell development under the condition of allelic inclusion. Immunity 6(3):225–233

20. Axelrod D (1981) Cell-substrate contacts illuminated by total internal reflection fluorescence. J Cell Biol 89(1):141–145

21. Axelrod D (1989) Total internal reflection fluorescence microscopy. Methods Cell Biol 30:245–270

22. Axelrod D (2001) Total internal reflection fluorescence microscopy in cell biology. Traffic 2(11):764–774

23. Axelrod D (2003) Total internal reflection fluorescence microscopy in cell biology. Methods Enzymol 361:1–33

24. Axelrod D (2008) Chapter 7: Total internal reflection fluorescence microscopy. Methods Cell Biol 89:169–221

25. Tolar P, Hanna J, Krueger PD, Pierce SK (2009) The constant region of the membrane immunoglobulin mediates B cell-receptor

clustering and signaling in response to membrane antigens. Immunity 30(1):44–55

26. Plochberger B et al (2010) Cholesterol slows down the lateral mobility of an oxidized phospholipid in a supported lipid bilayer. Langmuir 26(22):17322–17329

27. Baldwin T et al (2007) Wound healing response is a major contributor to the severity of cutaneous leishmaniasis in the ear model of infection. Parasite Immunol 29(10):501–513

28. Elso CM et al (2003) Leishmaniasis host response loci (lmr1-3) modify disease severity through a Th1//Th2-independent pathway. Genes Immun 5(2):93–100

Chapter 17

Use of Streptolysin O-Induced Membrane Damage as a Method of Studying the Function of Lipid Rafts During B Cell Activation

Heather Miller and Wenxia Song

Abstract

B-lymphocytes have the ability to repair their plasma membranes following injury, such as by bacterial cholesterol-dependent cytolysins. The repair process includes the removal of the pore from the inflicted region of the plasma membrane via lipid raft-mediated internalization. Lipid rafts are critical for B cell receptor (BCR) activation. Cholesterol-dependent pore forming bacterial toxins provide a useful tool for examining the role of lipid rafts in B cell activation and the underlying cellular mechanisms. This method serves as a great alternative of known cholesterol disruption reagents such as filipin, nystatin, and methyl-β-cyclodextrin. Here, we describe a method of damaging primary murine B cell plasma membranes with the *Streptococcus pyogenes* cytolysin, Streptolysin O (SLO), and monitoring levels of damage, repair and BCR activation.

Key words Membrane damage, Membrane repair, Lipid rafts, Streptolysin O (SLO), B cell activation

1 Introduction

Streptolysin O (SLO), produced by *Streptococcus pyogenes*, is a cytolysin that binds cholesterol to form 30 nm diameter pores in plasma membranes [1, 2]. SLO damage has been used for studying membrane repair in fibroblasts, epithelial and muscle cells. Similar to these cell types, B cells repair SLO-induced membrane damage through a Ca^{2+}-dependent lysosome exocytosis. Upon lysosome fusion with the plasma membrane, the lysosomal content is released, including acid sphingomyelinase (ASM). ASM generates ceramides, which elicits lipid raft-dependent internalization of the damaged membrane [3–6] (Fig. 1). Consequently, membrane wound repair leads to increased internalization of lipid rafts from the plasma membrane. When B cells are activated via the B cell receptor (BCR) during repairing SLO-elicited membrane damage, BCRs and wounded membranes compete for lipid rafts for endocytosis and are internalized into separate compartments [7].

Chaohong Liu (ed.), *B Cell Receptor Signaling: Methods and Protocols*, Methods in Molecular Biology, vol. 1707, https://doi.org/10.1007/978-1-4939-7474-0_17, © Springer Science+Business Media, LLC 2018

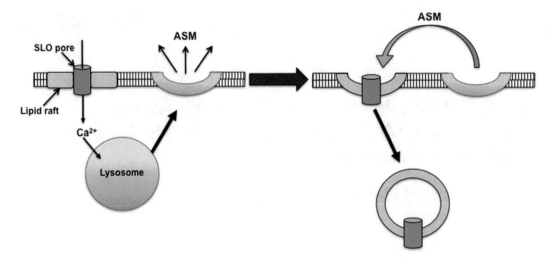

Fig. 1 SLO-elicited damage of B cell membranes induces lipid raft endocytosis. SLO binds to cholesterol to form pores, which allows Ca^{2+} to enter the cytoplasm and induce exocytosis of lysosomes. Lysosomes that fuse with the plasma membrane release acid sphingomyelinase (ASM) that generates ceramides, leading to internalization of the damaged membrane via lipid rafts

This membrane wounding-induced internalization of lipid rafts differs from other lipid raft disrupting reagents, such as filipin and nystatin that sequester cholesterol on the plasma membrane and methyl-β-cyclodextrin (MβDC) that extracts cholesterol from membranes [8, 9]. Moreover, membrane wounding by a bacterial toxin is a more physiologically relevant method of lipid raft disruption, as B cells are likely to encounter bacterial toxins at sites of infection.

Here, we describe methods of using SLO to damage primary murine B cell membranes and monitoring membrane damage, repair, and their effects on BCR activation using flow cytometry and immunofluorescence microscopy. The process involves using propidium iodide (PI) to detect cells with damage in their plasma membrane and immunofluorescence staining for phosphotyrosine as a means of detecting BCR activation. This method may also be applied for exploring the effects of lipid raft competition on the activation of other signaling molecules in BCR activation pathways. Furthermore, this technique can be applied for testing other toxins to determine their damaging potential on cell membranes and their effects on BCR signaling and endocytosis.

2   Materials

Any method of choice for isolating B cells from mouse spleens may be used. A general protocol for enrichment of B cells is described below.

2.1 Isolation of B Cells from Mouse Spleens

1. Frosted slides for macerating spleen.

2. Ficoll PM400 (Sigma or GE Lifesciences) for density-gradient centrifugation to enrich mononuclear cells and remove red blood cells.

3. Rat anti-mouse Thy1.2 mAb (BD Biosciences) and guinea pig complement (Rockland Immunochemicals) for T cell depletion. This incubation with the antibody and complement is performed in HBSS.

4. Final step will require panning in T-75 tissue culture treated flask to allow for monocytes and dendritic cells to attach.

2.2 SLO Damage of Cells

1. DMEM and Ca^{2+}-free DMEM, the latter of which is used as the media of no membrane repair control cells. In the absence of Ca^{2+}, membrane wounding will not induce lysosome exocytosis, consequently inhibiting wound repair. As the components of serum may interfere with SLO binding to the cells, only serum-free media, like DMEM, are used to incubate the cells with SLO.

2. Depending on the SLO used, it may be necessary to activate SLO using a reducing agent before use. SLO is oxidized in solution and loses activity. Sigma-Aldrich recommends a treatment of 20 mM cysteine or 10 mM dithiolthreitol. A mutated SLO, where a cysteine is deleted (provided by Rodney Tweten at University of Oklahoma), does not require this activation step.

3. Propidium Iodide (PI) for staining damaged and unrepaired cells.

2.3 BCR Activation and Phospotyrosine Staining

1. For the activation of BCRs, fluorescently conjugated F(ab')$_2$ anti-mouse IgM or IgM + IgG (Invitrogen or Jackson ImmunoResearch) may be used depending on the type of B cells.

2. 4% paraformaldehyde for fixing cells.

3. 0.05% saponin solution with 1% BSA for permeabilization/blocking.

4. Mouse IgG$_{2b}$ anti-phosphotyrosine mAb (4G10) (Millipore) and fluorescently conjugated goat anti-mouse IgG$_{2b}$ as a secondary antibody (Invitrogen) for analyzing BCR activation.

3 Methods

3.1 Isolation of B Cells from Mouse Spleens

1. Macerate mouse spleen with frosted slides and perform a density gradient centrifugation with Ficoll to isolate mononuclear cells. Centrifuge at $500 \times g$ for 30 min at 20 °C.

2. Suck off the buffy coat layer, obtain a cell count, and suspend cells at 1×10^7 cells/ml in HBSS.

3. Concentrations of Rat anti-mouse Thy1.2 mAb and guinea pig complement should be tested and optimized, typically 2–3 μg/ml of mAb and 1:10–1:20 dilution for the complement works well. Incubate the tube at 37 °C for 30 min, inverting the tube to mix every 10 min.

4. Wash the cells to remove antibody and complement and then incubate at 37 °C for 1 h in a T-75 flask in DMEM with 10% FBS, this will allow monocytes and dendritic cells to attach to the bottom of the flask while B cells will remain in suspension.

5. Pour off media and divide B cells into two separate tubes and wash the cells in one tube in DMEM without calcium (unrepaired control) and the other in DMEM with calcium.

3.2 SLO Wounding of B Cell Plasma Membrane

Optimization will be required for determining the concentration of cells and SLO. The optimization uses 5×10^6 B cells/ml with increasing concentrations of SLO, from 100 ng/ml to 400 ng/ml. If the lowest concentration of SLO wounds all B cells, increase the concentration of cells used.

1. Incubate 500 μl of optimized cell concentration/tube on ice with appropriate concentrations of SLO for 5 min to allow SLO to bind to the cells.

2. Warm the cells to 37 °C for 5 min, then immediately place the cells on ice and add propidium iodide (PI) to each tube.

3. Analyze using a flow cytometer immediately after to quantify the percentage of PI positive (PI+cells).

4. Determine the percentage of PI+ cells that have been incubated with DMEM $+Ca^{2+}$ as SLO wounded and unrepaired cells and the percentage of PI+ cells that have been incubated with Ca^{2+}-free DMEM as the total number of SLO wounded cells (Fig. 2). The percentage of repaired cells = $[\%PI+(-Ca^{2+}) - \%PI+(+Ca^{2+})] \times 100 / \%PI+(-Ca^{2+})$.

3.3 BCR Activation and Phospotyrosine Staining

Described here is immunofluorescent staining for total tyrosine phosphorylation. Anti-phosphotyrosine antibody can be substituted with an antibody of choice depending on which signaling molecule you are assaying for activation.

1. To activate BCRs, 500 μl of cells is incubated with fluorescently conjugated F(ab)$_2$_anti-mouse IgM or anti-mouse IgM + IgG (5–10 ug/ml) at 4 °C for 30 min.

2. The cells are washed to remove unbound antibody at 4 °C.

3. The cells are incubated with SLO at 4 °C for 5 min.

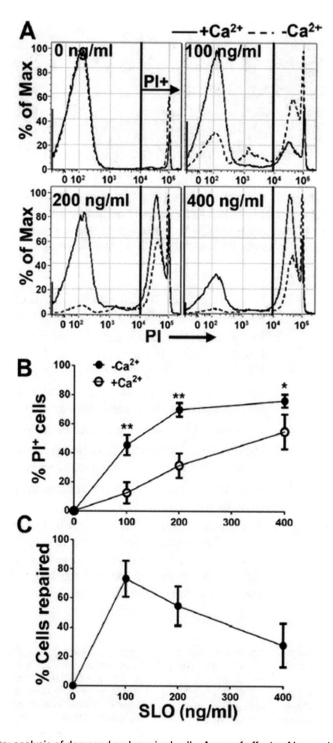

Fig. 2 Flow cytometry analysis of damaged and repaired cells. Assay of effects of increasing concentrations of SLO on B cell membrane repair. Without calcium ($-Ca^{2+}$) the cells do not repair, this serves as a control for determining how many cells were damaged by the SLO. With calcium ($+Ca^{2+}$) the cells are able to repair as can be seen by the decreased percentage of propidium iodide positive cells (PI+) compared to the cells without calcium (**a, b**). The percentage of repair for the different concentrations of SLO can be determined by [%PI positive ($-Ca^{2+}$)−%PI positive ($+Ca^{2+}$)] × 100/%PI positive ($-Ca^{2+}$) (**c**)

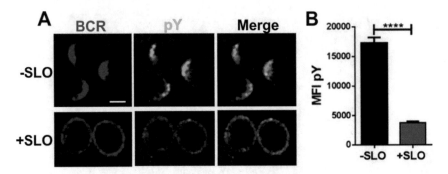

Fig. 3 Examination of effects of total BCR activation during SLO damage. BCRs were activated with a BCR cross-linking antibody and damaged with SLO for 5 min, then fixed and stained for phosphotyrosine (pY). The cells were imaged with a fluorescence confocal microscope (**a**) to examine distribution of phosphorylated molecules and run on a flow cytometer (**b**) to quantify levels of phosphorylation

4. Warm the cells to 37 °C for desired lengths of time and then fix with 4% PFA for 20 min at room temperature.

5. Permeabilize and block the cells with the permeabilization/blocking buffer (0.05% saponin and 1% BSA in PBS) for 30 min.

6. Stain with anti-phosphotyrosine antibody (1 μg/ml) in the permeabilization/blocking buffer for 30 min.

7. Wash three times with the permeabilization/blocking buffer.

8. Incubate the cells with fluorescently conjugated secondary antibody in the permeabilization/blocking buffer for 30 min.

9. Wash the cells with the permeabilization/blocking buffer for three times.

10. Analyze on a flow cytometer and image using a fluorescence confocal microscope (Fig. 3). For imaging, add ~250 μl of cells to a well of an 8-well chambered coverglass (Nunc Lab-Tek), the cells sit at the bottom of the well and image well on a fluorescence confocal microscope.

References

1. Bhakdi S, Tranum-Jensen J, Sziegoleit A (1985) Mechanism of membrane damage by streptolysin-O. Infect Immun 47:52–60

2. Los FC, Randis TM, Aroian RV, Ratner AJ (2013) Role of pore-forming toxins in bacterial infectious diseases. Microbiol Mol Biol Rev 77:173–207

3. Idone V, Tam C, Goss J, Toomre D, Pypaert M, Andrews N (2008) Repair of injured plasma membrane by rapid Ca^{2+} dependent endocytosis. J Cell Biol 180:905–914

4. Tam C, Idone V, Devlin C, Fernandes M, Flannery A, He X, Schuchman E, Tabas I, Andrews N (2010) Exocytosis of acid sphingomyelinase by wounded cells promotes endocytosis and plasma membrane repair. J Cell Biol 189:1027–1038

5. Corrottee M, Fernandes M, Tam C, Andrews N (2012) Toxin pores endocytosed during plasma membrane repair traffic into the lumen of MVBs for degradation. Traffic 13:483–494

6. Corrotte M, Almeida P, Tam C, Castro-Gomes-T, Fernandes M, Millis B, Cortez M, Miller H, Song W, Maugel T, Andrews N (2013) Caveolae internalization repairs wounded cells and muscle fibers. eLife 2:E00926

7. Miller H, Castro-Gomes T, Corrotte M, Tam C, Maugel T, Andrews N, Song W (2015) Lipid raft-dependent plasms membrane repair interferes with the activation of B-lymphocytes. J Cell Biol 211:1193–1205

8. Simons K, Toomre D (2000) Lipid rafts and signal transduction. Nat Rev Mol Cell Biol 1:31–39

9. Awasthi-Kalia M, Schnetkamp PP, Deans JP (2001) Differential effects of filipin and methyl-beta-cyclodextrin on B cell receptor signaling. Biochem Biophys Res Commun 287:77–82

Chapter 18

Visualization and Quantitative Analysis of the Actin Cytoskeleton Upon B Cell Activation

Vid Šuštar, Marika Vainio, and Pieta K. Mattila

Abstract

The formation of the immunological synapse upon B cell activation critically depends on the rearrangement of the submembranous actin cytoskeleton. Polymerization of actin monomers into filaments provides the force required for B cell spreading on the antigen-presenting cell (APC). Interestingly, the actin network also participates in cellular signaling at multiple levels. Fluorescence microscopy plays a critical role in furthering our understanding of the various functions of the cytoskeleton, and has become an important tool in the studies on B cell activation. The actin cytoskeleton can be tracked in live cells with various fluorescent probes binding to actin, or in fixed cells typically with phalloidin staining. Here, we present the usage of TIRF microscopy and an image analysis workflow for studying the overall density and organization of the actin network upon B cell spreading on antigen-coated glass, a widely used model system for the formation of the immunological synapse.

Key words Lymphocytes, B cells, Immunological synapse, Actin cytoskeleton, Microscopy, Total internal reflection microscopy, TIRF, Live cell imaging, Signaling microclusters, Image analysis

1 Introduction

The actin cytoskeleton is a pivotal player in lymphocyte activation as it provides the forces responsible for the dramatic change of cell morphology upon the formation of the immunological synapse (IS) [1–3]. Moreover, the actin cytoskeleton has been implicated in the regulation of cell signaling through the control of membrane protein diffusion and organization [4]. For example, the actin cytoskeleton affects B cell receptor (BCR) signaling by controlling interactions between the BCR and its co-receptors [5–7]. The importance of actin in the immune system is further demonstrated by diseases, such as the Wiscott-Aldrich syndrome (WAS), caused by defected action of an immune cell-specific actin-regulatory protein WASP [8, 9].

Within the first minutes of the B cell encounter of surface-bound antigens on antigen-presenting cells (APC), the cortical

Chaohong Liu (ed.), *B Cell Receptor Signaling: Methods and Protocols*, Methods in Molecular Biology, vol. 1707, https://doi.org/10.1007/978-1-4939-7474-0_18, © Springer Science+Business Media, LLC 2018

actin filament (F-actin) network is destabilized to promote efficient reorganization of the actin architectures [10]. Similarly to T cell IS, B cells form a radially protruding leading edge, composed of lamellipodia and filopodia [1]. This leads to the B cell spreading over the APC, critical for the subsequent gathering and internalization of antigen [2, 3].

The important role of actin in the IS is reflected by the distinct actin structures at different parts of the IS. Outermost lamellipodia consists of dense, highly branched actin meshwork that is followed by more aligned, concentric F-actin in the lamella and, finally, the central area largely void of obvious actin architectures [1]. Traditionally, in fixed cells the actin cytoskeleton has been visualized by staining with fluorescently labeled phalloidin [11]. With its high specificity for F-actin, phalloidin has maintained its position as the gold standard of actin staining. However, cell fixation and permeabilization with detergents can expose the samples for artifacts. Furthermore, the knowledge of the underlying dynamics of the system can only be achieved by visualization of the actin structures in living cells. Coinciding with the escalation of the usage of fluorescence microscopy and the development of fluorescent proteins in the last 10–15 years, also various actin tracers optimized for live cell microscopy have been introduced.

The first marker for live cell imaging of actin was GFP-actin [12]. While this probe is still highly relevant for applications such as FRAP (fluorescence recovery after photobleaching), its impaired competence in F-actin polymerization, potentially leading to incomplete, or biased representation of the cellular actin network, has raised the need for better markers [13]. Today, researchers can choose the most suitable tool for their studies from an array of actin tracers with different strengths and disadvantages. Most of these tracers are based on fusing fluorescent proteins (FP) to actin-binding peptides from different cytoskeletal regulators. When choosing the marker, one should take into account, e.g., the estimated coverage of different actin structures, potential influence on actin dynamics, as well as the probe kinetics informing about the probe suitability to the application of interest [14]. Currently, one of the most widely used actin tracers is LifeAct, a short actin-binding peptide fused to different FPs [15]. A relatively recent alternative is to use nanobodies, such as Actin-Chromobody (ChromoTek), utilizing actin-directed single-chain antibodies, fused to FP. One of the drawbacks of all these tracers is that they have to be transfected or genetically engineered into the cells of interest. To combat this issue, a cell-penetrable actin tracer, SiR-Actin (Spirochrome), was recently brought to markets.

BCR signaling and the subsequent remodeling of the actin cytoskeleton to form the IS occur at the close proximity to the plasma membrane. There are many microscopy techniques available to study lymphocyte activation on different scales, each with their own benefits and drawbacks [16]. Among those, the total internal reflection fluorescence (TIRF) microscopy represents a relatively easy-to-use, high-resolution method that is well suited for IS studies. In the Z-axis, TIRF reaches a resolution of 150–200 nm, which is significantly better than confocal microscopy with a typical resolution of ≥500 nm. High axial resolution in TIRF is achieved by the illumination of the cover glass–sample interface with a laser at high angle to cause total reflection of the incident light [17]. Consequently, the bottom of the sample is illuminated by an exponentially decaying evanescent field, which allows clear visualization of the phenomena at the cell membrane without interfering background fluorescence from the other parts of the cell. Very low total exposure to lasers minimizes the phototoxicity and makes TIRF particularly suitable for live cell imaging. Due to the sensitivity of the laser illumination, however, obtaining equal excitation intensity throughout the field of view can be challenging. This problem is effectively solved by ring-TIRF, a modern technique that fast-spins the laser light around the objective thereby creating an even TIRF illumination throughout the field of view [18].

In this article, we apply TIRF microscopy to visualize the actin cytoskeleton upon B cell activation by surface-bound antigens. We include an optimized cost-efficient protocol for transfection of B cell lines and describe the details of sample preparation for TIRF imaging of both fixed and living cells. Finally, we present an efficient image analysis work-flow to quantify cell spreading and actin intensity together with a threshold analysis of high-intensity structures in multiple channels. To provide complementary information about BCR signaling, the methods are compatible with addition of other markers, such as FP-Syk for live cell imaging, or phosphotyrosine (pTyr) staining in fixed samples, as shown in our examples.

2 Materials

2.1 Cells

1. BJAB human Burkitt lymphoma cell line, and A20 mouse B cell lymphoma line stably expressing the transgenic IgM BCR D1.3 [19] (*see* **Notes 1** and **2**).

2. Growth medium for BJAB and A20 cells: RPMI 1640 + 2.05 mM L-glutamine supplemented with 10% fetal calf serum (FCS), 50 μM β-mercaptoethanol, 4 mM L-glutamine, 10 mM HEPES and 100 U/ml Penicillin-Streptomycin.

2.2 Visualizing Actin Dynamics in Living Cells

2.2.1 Transfections

1. AMAXA electroporation (Biosystems).

2. 0.2 cm gap width electroporation cuvettes (Sigma-Aldrich®).

3. Recovery medium: growth medium supplemented with extra 10% FCS (total final concentration of FCS 20%) and 1% DMSO.

4. 2S transfection buffer: 5 mM KCl, 15 mM MgCl$_2$, 15 mM HEPES, 50 mM Sodium Succinate, 180 mM Na$_2$HPO$_4$/ NaH$_2$PO$_4$ pH 7.2 (*see* **Note 3**).

5. Pasteur pipettes fitting 2 mm gap width cuvette.

6. 6-well plates, cell culture quality.

7. Non-linearized plasmids encoding the markers of interest. For example, a fluorescent actin probe LifeAct-FP, together with a fluorescently tagged protein downstream of BCR signaling, such as Syk (*see* **Notes 4** and **5**).

2.2.2 Microscopy

1. 35 mm glass-bottom dishes (MatTek Corporation, Massachusetts, USA) (*see* **Note 6**).

2. Anti-BCR antibodies for BCR signaling-mediated cell spreading: anti-human IgM for BJAB, and anti-mouse IgM for A20 D1.3 (*see* **Note 7**).

3. Fibronectin for non-BCR-mediated cell attachment (*see* **Notes 8** and **9**).

4. Imaging buffer: 0.5 mM CaCl$_2$, 2 mM MgCl$_2$, 1 g/l D-glucose, 0.5% FCS in PBS.

5. An inverted fluorescent microscope with a TIRF module to acquire the images. Many vendors provide suitable systems. We have used either Zeiss Axio Observer Z1 with the setup briefly described below, or alternatively, DeltaVision OMX v4 (GE Healthcare Sciences) system with similar specifications, including ring TIRF, multiple scientific CMOS cameras, and lasers.

   (a) EMCCD camera (Hamamatsu, ImagEM, Model C9100–13).

   (b) Objective (Zeiss Alpha Plan-Apochromat 63×/1.46 NA, Oil Korr TIRF).

   (c) Incubator (XLmulti S1 DARK LS environmental chamber).

   (d) Multi laser module (laser lines 488, 561, and 640 nm).

   (e) Emission filters (Zeiss Filter set 77 HE GFP + mRFP + Alexa 633).

   (f) Vibration isolation table (Newport, Integrity 3 VCS Table System).

   (g) Image acquisition software (Zeiss ZEN Blue ver. 2).

2.3 Visualizing the Actin Cytoskeleton in Fixed Cells

1. Silicone gaskets (MultiWell Chamber coverslip, Grace Bio-Labs, Oregon, USA) and TIRF compatible coverslips fitting the size of the silicone gasket (for example: 24 mm × 40 mm, 0.17 mm thickness).

2. Anti-BCR antibodies or fibronectin for functionalizing the glass (*see* Subheading 2.2.2 and **Notes 7** and **9**).

3. Fixation buffer: 4% PFA in PBS.

4. Permeabilization buffer: 0.1% Triton-X in PBS.

5. Blocking buffer: 10 mg/ml BSA in PBS.

6. Antibodies and phalloidin.

    (a) Alexa Fluor® 555 Phalloidin (150 U/ml, Life Technologies, OR, USA). Prior to use, dilute 1:50–1:200 in blocking buffer.

    (b) Anti-pTyr Antibody (1 mg/ml, clone 4G10, Merck Millipore, Massachusetts, USA). Prior to use, dilute 1:500 in blocking buffer.

    (c) FITC Rat Anti-mouse IgG2b (0.5 mg/ml, clone R12–3, BD Sciences, California, USA). Prior to use, dilute 1:600 in blocking buffer.

7. PBS.

3 Methods

3.1 Visualizing Actin Dynamics in Living Cells

3.1.1 Transfections

1. Use 4 ml of recovery medium per well in 6-well plates. Warm up the recovery medium in an incubator (+37 °C, 5% CO_2) (*see* **Note 10**).

2. Centrifuge 4×10^6 cells down very gently (*see* **Note 11**).

3. Gently resuspend the cells in 180 μl of 2S transfection buffer containing up to 4 μg of non-linearized plasmid DNA (*see* **Note 12**). Place briefly on ice.

4. Transfer the cells with plasmid DNA to an ice-cold cuvette precooled on ice (*see* **Note 13**).

5. Transfect the cells using Amaxa nucleofector: for BJABs use program X-001 and for A20 cells, use program X-005.

6. Immediately after nucleofector pulse, fill the cuvette with warm recovery medium and transfer the cells into 6-well plates containing warm recovery media. Rinse the cuvette with recovery medium to collect the remaining cells.

7. Place the cells to the incubator to recover for 24 h (*see* **Note 14**).

8. Use flow cytometry to determine the transfection efficiency and viability of the cells (Fig. 1).

A

Cells (Amaxa program)	Viability	Expression (LifeAct)
A20 D1.3 (X-001)	75.34%	86.35%
BJAB (X-005)	75.56%	31.12%

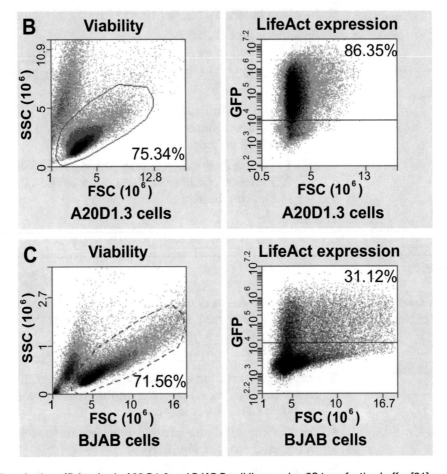

Fig. 1 Transfection efficiencies in A20 D1.3 and BJAB B cell lines, using 2S transfection buffer [21] and Amaxa nucleofector programs X-005 and X-001, respectively (**a**). 24 h after the transfection, 86% of A20 D1.3 (**b**), and 31% of BJAB (**c**) expressed LifeAct-GFP

3.1.2 Preparing the Microscope Slides

1. Functionalize the MatTek dishes as below. Keep in PBS until the cells are added. Do not let the dishes get dry.

   (a) Anti-IgM coating for BCR-mediated spreading: Use anti-IgM in the final concentration of 10 μg/ml in PBS. Incubate for 40 min at +37 °C. Rinse with PBS (*see* **Note 7**).

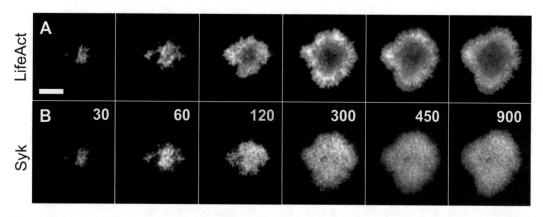

Fig. 2 Time-lapse image series of a BJAB cell on anti-human IgM-coated glass, imaged by TIRF microscopy. (a) LifeAct-Ruby as a marker for the actin cytoskeleton. (b) Syk-GFP as a marker for BCR signaling. The time since contact in seconds is shown in the upper right corner of (b) (*see* **Note 28**). Scale bar, 10 μm

(b) Fibronectin coating for adhering resting cells: Use fibronectin in the final concentration of 4 μg/ml in PBS. Incubate for 30–40 min at RT or +37 °C. Rinse couple of times with PBS.

3.1.3   Live Cell Imaging

1. Equilibrate the microscope by pre-heating it up to +37 °C (*see* **Note 15**).

2. Place the functionalized MatTek dish on TIRF objective to pre-warm for 5 min (*see* **Note 16**).

3. Carefully replace the PBS on the MatTek dish with 100,000–500,000 cells in 100 μl of imaging buffer (*see* **Note 17**). Note the time.

4. Focus on the cells, verify optimal TIRF illumination, and start imaging (Fig. 2) (*see* **Note 18**).

3.2   Visualizing the Actin Cytoskeleton in Fixed Cells

Below, we provide a protocol for immunofluorescence staining, adjusted for TIRF imaging, which uses reusable 8-well silicone gaskets as an economical alternative to perform the staining on MatTek dishes with considerably higher reagent volume requirements.

3.2.1   Preparing the Immunofluorescence Samples for TIRF Imaging

1. Place a silicone gasket on a cleaned (*see* **Note 19**) long TIRF quality coverslip (*see* **Note 20**) by gently pressing it on the glass (*see* **Note 21**).

2. Functionalize the slides as in Subheading 3.1.2 using the volume of 30 μl per well.

3. Apply 50,000–100,000 cells per well in 30 μl of imaging buffer (*see* **Note 17b**).

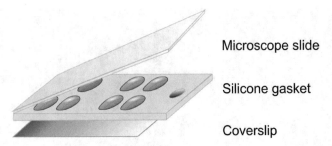

Microscope slide

Silicone gasket

Coverslip

Fig. 3 Assembly of the cover slip, silicone gasket, and the microscope slide for TIRF microscopy of fixed samples. The setup allows for low reagent consumption as compared to the usage of MatTek dishes, for example. In the final step, starting from one side, the microscope slide is gently pressed on the slightly overfilled sample wells to squeeze any air and excessive fluid out (*see* **Note 25**)

4. Activate the cells for the time of interest in the incubator at +37 °C (*see* **Note 22**).

5. Fix the cells for 10–15 min at the room temperature (RT). Rinse once with PBS (*see* **Note 23**).

6. Permeabilize the cells for 5 min, RT. Rinse once with PBS.

7. Block the cells for 1 h, RT.

8. Stain for F-actin and pTyr:

    (a) Apply fluorescently labeled phalloidin and anti-pTyr antibody to the cells (*see* **Note 24**). Incubate for 1 h, RT.

    (b) Rinse well (three times) with PBS.

    (c) Apply the secondary antibody against anti-pTyr to the cells. Incubate for 1 h, RT.

    (d) Rinse well (three times) with PBS and overfill the wells with PBS.

9. Cover the gasket with a clean (*see* **Note 19**) microscope slide to seal it, and to provide support for handling (Fig. 3) (*see* **Note 25**).

3.2.2 TIRF Imaging of Fixed Samples

As in Subheading 3.1.3, **step 4**. A notable difference can be seen on cells spread on anti-IgM versus fibronectin, visualized both by phalloidin and pTyr stainings (Fig. 4).

3.3 Data Analysis

Here, we provide an example of how to use the open source software Image J to quantitatively analyze the actin cytoskeleton, together with markers such as pTyr, during B cell activation. The same work-flow can be applied for analysis of both fixed or live cells. The surface area of B cells is quantified based on actin intensity. In addition, through local thresholding, we analyze actin-rich areas as well as pTyr-enriched puncta, indicating signaling microclusters. We present a series of steps, applicable in various image analysis software, required to obtain such quantitative information. These

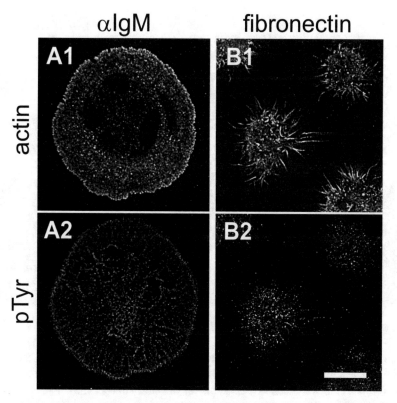

Fig. 4 BJAB cells were let to spread for 20 min at 37 °C on cover slips coated with 10 µg/ml anti-human IgM (**a1-2**) or 4 µg/ml Fibronectin (**b1-2**). The cells were fixed and stained with phalloidin AlexaFluor-555 (**a1, b1**) and anti-pTyr antibodies followed by secondary antibodies labeled with AlexaFluor-488 (**a2, b2**). The images were deconvoluted (*see* **Note 28).** Scale bar, 10 µm

steps can be combined in a script for semi-automatized batch analysis, such as the script we have made available, together with informative sample images, at our website (http://mattilalab.utu. fi/). For more detailed information about actin dynamics and retrograde flow live cell imaging together with specialized analysis methods is required. For example, a method for analyzing the actin retrograde flow in immune synapse through time-lapse kymographs is shown in [20].

1. Import raw images to Image J (Bioformats plugin).

2. Select the regions of individual cells on the actin channel and make three duplicates of each selection (Fig. 5a1, d1) (*see* **Note 26**).

3. Restore the selections on pTyr channel and make two duplicates.

4. In order to measure the total surface contact area, perform automatic thresholding (Huang method) on one duplicate on actin channel. On the binary image, fill holes in order to get a single area.

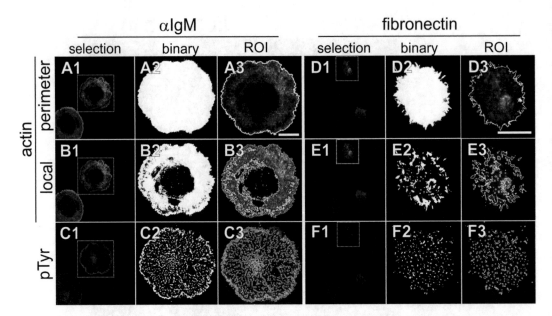

Fig. 5 Image analysis work-flow. BJAB cells were let to spread on cover slips coated with 10 μg/ml anti-human IgM (**a–c**) or 4 μg/ml Fibronectin (**d–f**) and prepared for imaging as in Fig. 4. The cells were stained with phalloidin (**a, b, d, e**) and anti-pTyr antibodies (**c, f**). Columns on the left (**a1–f1**) show the selection of individual cells in actin and pTyr channels. The middle columns (**a2–f2**) show automatic thresholding of whole cell perimeter (**a2, d2**) or local structures (**b2, c2, e2, f2**) into binary image. The columns on the right (**a3–f3**) show the ROIs, obtained from the binary image in the middle column, on the original image for the measurements of parameters of choice (area, intensity etc.). Scale bar, 10 μm

5. Create automatic selection based on thresholded binary image (Fig. 5a2, d2).

6. Restore the thresholded selection on non-thresholded duplicate of the same cell on actin channel, to create the region-of-interest (ROI) to be measured (Fig. 5a3, d3).

7. Select the parameters of choice for measurement, such as area, perimeter, and various intensity parameters.

8. Measure within the ROI and save the measurements.

9. In order to measure only the regions enriched in actin, perform local thresholding (Bernsen method) on actin channel on second original duplicate, to create a local thresholded image (Fig. 5b2 and e2). Specify the thresholding radius based on the minimum size of the structures you are interested in (*see* **Note 27**).

10. Repeat the creation of ROI, now based on local thresholding (Fig. 5b3, e3), as in **steps 5–8**.

11. Measure the pTyr puncta, by repeating **steps 9–10** for pTyr channel duplicates, using adjusted radius for local thresholding (Fig. 5c1–3, f1–3).

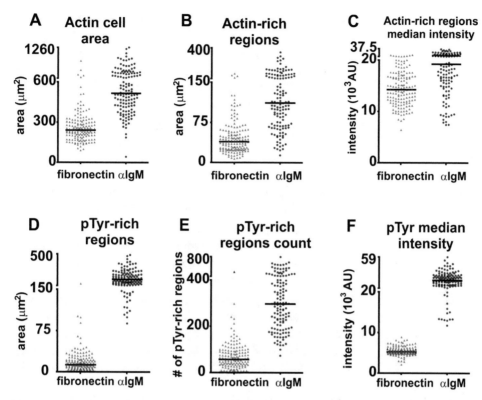

Fig. 6 Measurements obtained from an experiment illustrated in Fig. 5. (**a**) Whole cell area as thresholded on actin image, as shown in Fig. 5a2 and d2. (**b**) Total area of the actin-rich regions obtained via local thresholding, as shown in Fig. 5b2 and e2. (**c**) Median intensity of the actin-rich regions, as shown in Fig. 5b3 and e3. (**d**) Total area of the pTyr-rich regions obtained via local thresholding, as shown in Fig. 5c2 and f2. (**e**) Number of pTyr-rich regions per cell as shown in Fig. 5c2 and f2. (**f**) Median intensity of the pTyr-rich regions, as shown in Fig. 5c3 and f3. The data was obtained using a script allowing for batch analysis using Image J, freely available at (http:/mattilalab.utu.fi/). Each *dot* represents one cell and the *black line* median value. 120 cells on anti-IgM coted cover slips and 150 cells on Fibronectin coated cover slips were analyzed

12. Automatize the steps from 3 to 11 in Image J, by recording the steps in the macro recorder and copy and generalize the used commands in Image J Scripting console.

13. Import the results to a data analysis program, like MS Excel or GraphPad Prism, for graphical presentation and calculation of average, standard deviation, and statistical significance (Fig. 6).

4 Notes

1. The cells should be cultured in the complete growth medium. Keep the cells in their logarithmic growth phase for the day of the experiment or transfection.

2. While A20 is a more common research tool, BJABs allow better visualization of the actin cytoskeleton, while they spread with broader lamellipodia (Fig. 4). BJABs are available through *Thermo Fisher Scientific* or *Deutsche Sammlung von Mikroorganismen und Zellkulturen GmbH* and A20 through *Thermo Fisher Scientific* or ATCC®, for example.

3. The 2S transfection buffer composition is published in [21], which we adjusted by a small modification in the concentration of Na_2HPO_4/NaH_2PO_4 (180 mM). Preparation: weigh all the components into a 100 ml flask. Add water up to 80–90 ml. Adjust pH with NaOH to 7.2. Complete with water up to 100 ml, sterile filter (0.2 pore size), and freeze in aliquots.

4. As an example, we used here GFP-Syk and Ruby-LifeAct. Plasmids encoding for various actin probes in different colors can be bought from Addgene (Cambridge, USA), as well as constructs for BCR signaling proteins.

5. In our experience, GFP–anti-actin nanobody/Actin-Chromobody behaves quite comparably to FP-LifeAct in B cells and shows similar transfection efficiencies. In our brief experience, SiR-Actin, the cell penetrable actin probe, does not reach to comparable coverage of the B cell actin cytoskeleton.

6. MatTek dishes can be purchased in two different qualities. TIRF quality with lower variability in cover slip thickness is recommended for TIRF imaging for best performance.

7. Choose the correct anti-BCR depending on the receptor expressed in your cells. We use here anti-human IgM for BJAB, and anti-mouse IgM for A20 D1.3.

8. Aliquot fibronectin in the concentration of 1 mg/ml and avoid re-freezing the aliquots.

9. Alternatively, poly-L-lysine and antibodies against irrelevant, or largely non-signaling receptors (e.g., 1 μg/ml of anti-MHCII M5/114 (ATCC TIB120) for mouse primary B cells as in [6]) can be used. However, it is important to consider that all these reagents are likely to affect the cells in some way and the concentrations should be optimized for each cell type.

10. Make sure that the recovery medium is warm when adding the cells (e.g., use a heat block under the plate). The temperature of the recovery medium will affect the transfection efficiency and cell viability.

11. Cell viability is increased by gentle handling. For example, centrifugation should be performed with minimum rpm and time (we routinely use 18.0G for 5 min). To reduce shear stress, use bigger, 1 ml tips when possible.

12. We normally use total of 4 μg of plasmid DNA for BJAB and 3 μg (or 4 μg in the case of double of transfection) for A20 cells. The total volume of DNA should be kept minimal, not more than 10 μl, to ensure successful electroporation pulse.

13. Pipette the cells carefully along the cuvette wall to avoid any air bubbles. Make sure that the cuvette is dry from outside.

14. The expression is typically visible already within a few hours of nucleofection and can last for 2–3 days. Longer recovery time increases the viability of the cells; however, it reduces the percentage of cells expressing the ectopic marker.

15. Make sure that you are adequately trained for the use of the TIRF microscope and familiar with the procedures how to adjust the laser angle and verify the optimum TIRF illumination for your samples.

16. Try to place the dish as straight as possible on the objective, as any tilt will affect the TIRF angle. Also, the more straight your dish is, the less focus correction is required when moving around the dish.

17. (a) You can pre-warm the cells briefly beforehand to minimize focal shifts due to the temperature changes. (b) If you wish to do image analysis for the area of spreading on anti-BCR, adjust the cell number down to avoid close contacts between the cells.

18. By switching among fluorescent channels check for possible chromatic aberration difference between the channels and apply focus correction to compensate it.

19. To enable cleaning of the cover slip and thereby more efficient coating, we recommend the removal of the silicone gasket from its original cover slip. Clean the cover slip, e.g., with isopropanol and ethanol.

20. You can remove the dust on the silicone gasket by pressing it with tape. For easier handling, you can trim the silicone gasket with scissors to fit it on 4 cm long coverslip.

21. Try to avoid having bubbles between the glass and silicone to ensure the gasket will not leak.

22. For longer time points usage of an incubator is recommended. However, for shorter time points, such as 5–15 min, a heat block or even incubation in room temperature, for possible practical reasons, can be sufficient.

23. Throughout the staining procedure, be gentle when aspirating and pipetting to avoid cell detachment.

24. Phalloidin is good to be added in the first staining step, as it also stabilizes the F-actin.

25. Prior to applying the microscope slide, overfill the wells with PBS with the help of surface tension to generate a drop higher than the gasket. Press a microscope slide on the top of the gasket starting from one side; by this you will remove the excess of PBS and avoid air bubbles. Press the slide on the gasket for a moment and dry with a tissue from the edges to ensure it is attached firmly.

26. When analyzing live samples, one must take into account the added variation due to possible variation in the expression levels.

27. The thresholding radius (in pixels) depends on the samples and the resolution of the images. The optimal value should be empirically determined.

28. Figure 2 image and images used for the quantitative analysis were acquired with Zeiss Axio Observer TIRF microscope and ZEN 2 software. The example images in Figs. 4 and 5 were acquired using DeltaVision OMX v4 (GE Healthcare Sciences) microscope using ring TIRF mode. Figure 4 image was deconvoluted with DeltaVision softWoRx software.

Acknowledgments

Imaging was performed at the Cell Imaging Core, Turku Centre for Biotechnology, University of Turku and Åbo Akademi University. We thank Markku Saari and Jouko Sandholm from the Cell Imaging Core for their help with the microscopes, and the whole Mattila lab for their critical comments to the manuscript. Turku University Hospital (TYKS) is acknowledged for the support in basic laboratory materials. This work was supported by the Academy of Finland (grant ID: 25700, 296684 and 307313 for PKM, and 286712 for VŠ), University of Turku Graduate School (UTUGS) (for MV), VSSHP research support (for PKM), as well as Sigrid Juselius (for PKM), Jane and Aatos Erkko (for PKM), Magnus Ehrnrooth (for PKM and VŠ) and Finnish Cultural (for VŠ) foundations.

References

1. Le Floc'h A, Huse M (2015) Molecular mechanisms and functional implications of polarized actin remodeling at the T cell immunological synapse. Cell Mol Life Sci 72:537–556. https://doi.org/10.1007/s00018-014-1760-7

2. Batista FD, Iber D, Neuberger MS (2001) B cells acquire antigen from target cells after synapse formation. Nature 411:489–494. https://doi.org/10.1038/35078099

3. Kuokkanen E, Šuštar V, Mattila PK (2015) Molecular control of B cell activation and

immunological synapse formation. Traffic 16:311–326. https://doi.org/10.1111/tra.12257

4. Mattila PK, Batista FD, Treanor B (2016) Dynamics of the actin cytoskeleton mediates receptor cross talk: an emerging concept in tuning receptor signaling. J Cell Biol 212:267–280. https://doi.org/10.1083/jcb.201504137

5. Gasparrini F, Feest C, Bruckbauer A, Mattila PK, Müller J, Nitschke L, Bray D, Batista FD (2015) Nanoscale organization and dynamics of the siglec CD22 cooperate with the cytoskeleton in restraining BCR signalling. EMBO J 35:1–23. 10.15252/embj.201593027

6. Mattila PK, Feest C, Depoil D, Treanor B, Montaner B, Otipoby KL, Carter R, Justement LB, Bruckbauer A, Batista FD (2013) The actin and tetraspanin networks organize receptor nanoclusters to regulate B cell receptor-mediated signaling. Immunity 38:461–474. https://doi.org/10.1016/j.immuni.2012.11.019

7. Treanor B, Depoil D, Bruckbauer A, Batista FD (2011) Dynamic cortical actin remodeling by ERM proteins controls BCR microcluster organization and integrity. J Exp Med 208:1055–1068. https://doi.org/10.1084/jem.20101125

8. Derry JM, Ochs HD, Francke U (1994) Isolation of a novel gene mutated in Wiskott-Aldrich syndrome. Cell 78:635–644. doi: 0092-8674(94)90528-2 [pii]

9. Machesky LM, Insall RH (1998) Scar1 and the related Wiskott–Aldrich syndrome protein, WASP, regulate the actin cytoskeleton through the Arp2/3 complex. Curr Biol 8:1347–1356. https://doi.org/10.1016/S0960-9822(98)00015-3

10. Freeman S, Lei V, Dang-Lawson M, Mizuno K, Roskelley CD, Gold MR (2011) Cofilin-mediated F-actin severing is regulated by the rap GTPase and controls the cytoskeletal dynamics that drive lymphocyte spreading and BCR microcluster formation. J Immunol 187:5887–5900. https://doi.org/10.4049/jimmunol.1102233

11. Wulf E, Deboben a BF a, Faulstich H, Wieland T (1979) Fluorescent phallotoxin, a tool for the visualization of cellular actin. Proc Natl Acad Sci U S A 76:4498–4502. https://doi.org/10.1073/pnas.76.9.4498

12. Ballestrem C, Wehrle-Haller B, B a I (1998) Actin dynamics in living mammalian cells. J Cell Sci 111(Pt 1):1649–1658

13. Melak M, Plessner M, Grosse R (2017) Actin visualization at a glance. J Cell Sci 130 (3):525–530. https://doi.org/10.1242/jcs.189068

14. Belin BJ, Goins LM, Mullins RD (2014) Comparative analysis of tools for live cell imaging of actin network architecture. BioArchitecture 4:189–202. https://doi.org/10.1080/19490992.2014.1047714

15. Riedl J, Crevenna AH, Kessenbrock K, Yu JH, Neukirchen D, Bista M, Bradke F, Jenne D, T a H, Werb Z, Sixt M, Wedlich-Soldner R (2008) Lifeact: a versatile marker to visualize F-actin. Nat Methods 5:605–607. https://doi.org/10.1038/nmeth.1220

16. Balagopalan L, Sherman E, V a B, Samelson LE (2011) Imaging techniques for assaying lymphocyte activation in action. Nat Rev Immunol 11:21–33. https://doi.org/10.1038/nri2903

17. Axelrod D (1981) Cell-substrate contacts illuminated by total-internal reflection fluorescence. J Cell Biol 89:141–145. https://doi.org/10.1083/jcb.89.1.141

18. Mattheyses AL, Shaw K, Axelrod D (2006) Effective elimination of laser interference fringing in fluorescence microscopy by spinning azimuthal incidence angle. Microsc Res Tech 69:642–647. https://doi.org/10.1002/jemt.20334

19. Williams GT, Peaker CJG, Patel KJ, Neuberger MS (1994) The α/β sheath and its cytoplasmic tyrosines are required for signaling by the B-cell antigen receptor but not for capping or for serine/threonine-kinase recruitment. Proc Natl Acad Sci U S A 91:474–478. https://doi.org/10.1073/pnas.91.2.474

20. Jankowska KI, Burkhardt JK (2017) The immune. Synapse 1584:7–29. https://doi.org/10.1007/978-1-4939-6881-7

21. Chicaybam L, Sodre AL, Curzio BA, Bonamino MH (2013) An efficient low cost method for gene transfer to T lymphocytes. PLoS One 8(3):e60298. https://doi.org/10.1371/journal.pone.0060298

INDEX

Chaohong Liu (ed.), *B Cell Receptor Signaling: Methods and Protocols*, Methods in Molecular Biology, vol. 1707,
https://doi.org/10.1007/978-1-4939-7474-0, © Springer Science+Business Media, LLC 2018

Printed in the United States
By Bookmasters